OIL ON THE EDGE

OIL ON THE EDGE

Offshore Development, Conflict, Gridlock

By
Robert Gramling

STATE UNIVERSITY OF NEW YORK PRESS

Published by
State University of New York Press, Albany

For information, address State University of New York Press,
State University Plaza, Albany, N.Y., 12246

Production by Cathleen Collins
Marketing by Theresa Abad Swierzowski

Library of Congress Cataloging in Publication Data

Gramling, Robert, 1943–
    Oil on the edge : offshore development, conflict, gridlock / by
Robert Gramling.
        p.   cm.
    Includes bibliographical references and index.
    ISBN 0-7914-2693-9. — ISBN 0-7914-2694-7
    1. Offshore oil industry—United States. 2. Offshore oil
industry—Government policy—United States. 3. Offshore oil
industry. I. Title.
    HD9565.G73   1996
    338.2'7282'0973—dc29   94-48605
                                                      CIP

10 9 8 7 6 5 4 3 2 1

To Eileen

Contents

List of Tables

List of Figures

Preface

As a child I spent my summers on the Atlantic coast of Florida south of Jacksonville. I became a perpetual beachcomber endlessly bringing back treasures for my mother to admire, thereby reinforcing an activity that has been a joy for a lifetime. I learned to fish from my father and have been a fisherman as long as I can remember. The Whiting and Red Drum of the Atlantic, sea turtles laying their eggs at night, the wide expanse of the white sand beach became a part of me, and though I did not have the concepts to express it, I came to understand that these things were beyond value in the traditional sense of the word. We moved to Tallahassee and my explorations moved to the Gulf of Mexico. I fished for bountiful Speckled Trout on the grass flats off St. Marks and Spanish Mackerel off Cape San Blas when the great schools seemed almost thick enough to walk on. In my own lifetime I have seen this abundance dangerously diminish.

I came to Louisiana over two decades ago and in time came to love the coastal marshes as much as the white sands of my youth. By the time I arrived in Louisiana, wherever I went in the coastal marshes and offshore I encountered the incongruous shapes of coastal and offshore oil development. Given my love of the coast it was impossible not to become curious about these anomalies as I fished around them in coastal waters and offshore.

I first began to seriously study the process of offshore development in 1976 and have been doing so, off and on, ever since. This book is not a simple catalogue of offshore development, but rather a reflection of what I think is important for us to know and remember in order to learn from our mistakes. At its broadest level, the evolution of offshore development provides both a microcosm of technological change and the subsequent impacts on our human and physical environment. It also provides an example of how various policy scenarios dictated at the highest levels of our government mitigated or exacerbated those impacts. In short, the offshore development scenario provides a laboratory for the study of policy and social impacts. This book is an attempt to use that laboratory.

The story of offshore development is one of almost unparalleled technological evolution, as the search for oil moved first into the coastal marshes, then the nearshore estuaries, and finally offshore. Today production platforms in the Gulf of Mexico stand in over thirteen hundred feet of water and over a hundred miles offshore, and there are over three thousand of them in federal waters alone. In order to accomplish this level of growth, however, a policy environment that encouraged development was necessary. The consequences of this rapid development, within an encouraging policy context and during an exuberant era with an almost unlimited faith in technology, were alterations of the coastal environment, the infrastructure, and human capital. The effects of these alterations are only beginning to be fully realized.

Today the majority of the offshore federal lands in the United States are closed by Presidential and Congressional action to the very agency within the Department of Interior that was created to lease and manage them. It would be difficult to interpret this as anything but a statement on past policy and management initiatives. The current gridlock on the Outer Continental Shelf serves no one, neither those who support nor those who oppose offshore development. Perhaps, by examining how we came to be at this impasse we can learn something about how to avoid similar situations in the future.

This research was funded in part by the Minerals Management Service, U.S. Department of Interior. The views, discussions, and conclusions in this book are, however, entirely those of the author.

Prologue

Traveling Through Time

Flying south from New Orleans by helicopter the terrain changes abruptly. The tall buildings of the central business district, standing sentinel along the meandering Mississippi, which to our left through its own serpentine mentality actually flows northward, soon fall behind. The river itself, home to half the waters of a country, rises anomalously over the level landscape. Centuries of impoundment, first by the early settlers and now the Corps of Engineers, and many more centuries of eroding, transporting, and storing the soils of North America in its bed have left the river perched above the surrounding plain. Temporarily captured between ever-growing levees, the tourist in Jackson Square can look upward at ships navigating the flow. Joining these tankers and freighters from the nations of the world, the sinuous lines of the banks are further scarred by warehouses; the drips of rusting, obsolete machinery; hundreds of barges in various states of disrepair; dilapidated but functional loading yards; and, upstream far in the distance, by the massive incongruous shapes and plumes of giant refineries. Passing over the river and the Canal Street ferry, chasing its dock crabwise against the current, an observant traveler might notice upstream the lock where vessels are raised up from sea level on the Gulf Intracoastal Waterway to the artificial elevation of the river. And then there is the urban sprawl of the west bank settlements flickering below.

Neighborhoods, shopping centers, streets, and boulevards buried under toy automobiles, junkyards, and azure swimming pools

1

all flash by. The terrain is level. It is a bright spring day, and in the distance ahead human settlement thins and a motif of brilliant greens interspersed with blues reaches away to the hazy horizon. A smaller levee than those that incarcerate the Mississippi separates the denser human living space from the encircling wetlands, and in a deceptively short time from leaving the trolleys plying St. Charles Avenue, we are immersed in the marsh.

To our left the Mississippi, the primeval power in the land we traverse, has once again established its mission, flowing sedately southeastward. Below, a flock of snow white egrets, sailing through the spring morning at crosspurposes to our direction, punctuate the emerald landscape. To our right a small blacktop road rolls south along the only solid land, creating bands of settlements along the natural levee of Bayou des Familles. The road will end at Lafitte, the community named after the colorful pirate who dominated this watery landscape well over a century ago. But our journey lies southwestward across the road, along the shores of Lake Salvador, and out over the deltaic plain.

This is a vast, open land laid down by the Mississippi River over millennia as it writhed across southern Louisiana, melding land and water in a never-ending succession of channelization, deposition, and erosion. It spreads from the Chandeleur Islands off the coast of Mississippi westward almost two hundred miles to Vermilion Bay. In places, the great liquid plain extends almost sixty miles inland from the shoreline, and even the casual observer can see that this is like no other place on the continent.

The trembling prairie below is too fragile to support the weight of a traveler, and movement across this land by humankind is limited to vessels on the numerous lakes and bayous or to the comparatively solid ground of the natural levees of bayous as they cross the wetlands pouring coastward. The environs of the enormous marsh are color coded as we proceed overhead. The brilliant greens of the new growth marsh grass stretch as the basic background on which the quilt of the more active wetland processes are painted. The natural levees of the bayous are washed with the mixture of complementary greens of new growth cypress and willow and the hazy grays of still deciduous hardwood, punctuated by the multihued cyan blooms of red maple and the great dark green clouds of live oak. The open water vacillates between the chocolate of the shallow lakes to the stained black of the deep bayous. From the air the sinewy weave is a color negative of a venous system, transformed from reds to greens.

Crossing Bayou Lafourche, we see the tenuous bargain between humanity and this vast aqueous landscape. Human settlement is compressed onto narrow ribbons of natural levee on either side of the waterway. These "string towns" miles long, but in places yards wide, were once linked only by water. Even now, in places, the road runs naked along the bayou, and though the precise, angular patterns of human habitation stretch laterally out of sight, they seem isolated along the narrow asphalt artery, huddled between the press of the open marsh.

We can measure the relationship of these people to this world by the genealogy of boats. Offshore shrimping vessels pass below, with their stately utilitarian lines and their massive flanking booms raised prayerfully overhead to navigate the confines of the bayou. More numerous are the smaller coastal shrimpers that work the estuaries and bayous of this world—with the square frames of their "butterfly" nets folded and gossamer in the sun, like a swarm of sturdy nautical insects poised for flight. Still more diminutive are the Lafitte Skiffs, uniquely Bayou Lafourche vessels, with their graceful lines and culling decks overhanging their sterns. Even in yards, on trailers, and on the highways towed behind pickup trucks, we see boats. Most of these vessels are harvesters, and many of the designs embody adaptations over generations. For not only is this a vast land, and a tenuous one for our species, but in antithesis to its austere appearance, it is also an incredibly abundant land.

Unlike many of our national resources—farms where soil is being used up faster than it can regenerate; virgin timber largely gone; coal mines of the West and Appalachia stripped and vacant; or even the once bountiful fishing banks of the Northeast coast, now overfished and in decline—this land, centered around the annual cycle of the shrimp, still manages to provide a harvest that renews itself each year. The thousands of square miles of marsh and estuaries support one of the most prolific food chains on earth. Speckled trout and redfish swim these waters, and shrimp in incredible numbers breed and proliferate. These are the nursery grounds for almost a quarter of the commercial fisheries of the United States, a resource of incomparable value.

We see other vessels on the bayou also. Less graceful, more utilitarian, and obviously more powerful, these boats service the coastal oil and gas industry. Ponderous supply vessels, with their elevated bows and low flat decks designed to carry massive weights, tower above fishing boats along the bayou. Sleek "crew" boats, still larger than the largest fishing vessel, plane over the surface of the water leaving behind imposing wakes that crash into the surrounding marsh. Strange, awkward bargelike crafts with spindly legs that lift them clear of the water, perch top heavily along the banks with all of the grace of elephants stilt-walking. These vessels seldom negotiate the marsh now on their way to oil and gas installations, since those reservoirs are mostly drained and abandoned. They live in the deeper bayous waiting to go offshore where we are headed.

There is another phenomenon that becomes increasingly evident as we move across this world. Many of the water bodies below, although bayoulike in their width, show a singularity of purpose as they roll across the landscape that means they can only be human in origin. Some travel out of sight in both directions in razor-straight lines. Others are systems of dead-end mazes with rectangular branches from a common origin. Within them, alien wooden structures perch on decaying pilings, scattered along these artificial waterways like pieces of piers that wandered away from harbor and became lost. Along their banks, fingers of erosion cut into the wetlands.

Like the bayous, these too are obviously travel routes, conforming not to the lay of the land, they were driven with brute force through the unresisting wetlands with an economy of goal-directed intent to do an accountant proud. For the most part these are old structures and canals. They represent perhaps the saddest time for this fragile estuarine ecosystem since the arrival of humans thousands of years ago, a time when the removal of oil from beneath the marsh was accomplished in the most direct and efficient manner, with little understanding of the value or the fragile nature of these wetlands. Because the marsh would not support the heavy equipment needed to drill for oil, it was dissected to allow barges with equipment to reach the well locations, and to lay the pipelines that carried the fruits of this venture to market. Since the logic of well locations and hence these canals exists only in obscure seismic evidence, they bear no rational relationship to the natural processes occurring on the land around them and live like scars on an otherwise coherent canvas. As with the strip mines in the eastern United States, the scars remain long after those responsible have taken their profit and are gone.

Nearing the coast the land-water ratio begins to tip in a liquid direction and the patterns of green give way to a multihued theme of blue, and even over the hot acid smell of aviation fuel we can began to smell the salt. More structures dot the open waters of Timbalier Bay and the adjoining estuaries; some are connected by complicated networks of catwalks, some stand alone. These platforms are also old. Battered pilings support weather-beaten decks and soft resilient gray buildings. But they are busy, with boats moored alongside some of the larger structures and evidence of human habitation on many of them. And then, we are at the Gulf of Mexico, stretching south 500 miles to the Yucatan Straits.

In this ever-eroding and accreting landscape it is often difficult to clearly mark a coastline. Here, however, Timbalier Island provides a rambling sliver of a barrier to the open Gulf, sheltering the bay and the oil installations behind it. On the island are more oil-related buildings and structures, partners to those in the bay, and then in an eerie Darwinian fashion, before our eyes, they creep offshore.

There are platforms in the Gulf barely beyond the surf, still clinging tenaciously to terra firma with spidery bridges to land. Beyond them, separated from the shore, are small, clearly aging structures, huddled together in clusters for protection and mutual support against the raw force of the open Gulf. Still further out, as far as the eye can see, of ever-growing size, are mammoth constructions of pipe and steel that have no progenitor on land. These production platforms are the bases from which the wealth of oil beneath the sea is harvested. As we move further offshore we pass over an invisible line, three miles from the coast, that differentiates the waters owned by the state of Louisiana from those federal lands belonging to the people of this entire country and entrusted to the Department of Interior.

The platforms grow larger and show less age, but they do not disappear. Never are we out of direct sight of dozens of platforms, and at times we can literally see hundreds stretching toward the horizon.

A line of black cotton thunderstorms comes into sight from the south and forces us landward, but it is evident that we have only surveyed the veneer of offshore development in the Gulf. Indeed, had we time, patience, and resources, we could count almost 4,000 of these platforms in federal waters alone. They are home to over 18,000 production wells and are connected to one another and to shore by a spider web of thousands of miles of undersea pipelines. In places they stand stoically over a hundred miles offshore in water depths of well over a thousand feet and represent perhaps the most massive extractive development scenario in human history and one of the most rapid adaptations of technology in modern times.

We flee the approaching line of storms backward through the evolution of offshore structures in a wide arc to the west and cross the coast near the mouth of the Atchafalaya River. Here the bay gives us one of the few examples of growth in this vanishing rich wilderness. With most of the valuable silt necessary to nourish these lands carried by the imprisoned Mississippi down its bird-foot delta and dropped unproductively off the continental shelf into the deep ocean basin, the Atchafalaya, a southwesterly flowing distributary and the primary overflow valve for the Mississippi, provides one of the few ways the replenishing silt can reach the shallow coastal shelf.

Following the river inland we soon come to the natural levee of Bayou Teche, the historic migration route into this region, pinched between the Atchafalaya River basin overflow swamp to the north and the ubiquitous coastal marsh. Again, human habitation and vessels of all types are abundant. Here the more recent Intracoastal Waterway bisects the Atchafalaya River, and water transportation is available to the four points of the compass. The towns of Patterson, Berwick, and Morgan City nestle along the water.

Once fishing communities and now support centers for the thousands of offshore platforms, their transformation is evident.

Turning east over Morgan City, paralleling the Intracoastal Waterway, the turn of the century downtown buildings and Victorian homes soon give way to concrete fields and shopping centers, which in turn are replaced by industrial sites along the waterway. Scrap yards with corroding hills of steel coexist with hundreds of acres of drill pipe and well casing, neatly laid out and rusting in the sun. Further on we cross the huge fabrication yards where most of the offshore structures we saw were fashioned. An offshore platform under construction lies prostrate, surrounded by spindly cranes, like a flock of hungry herons contesting a particularly delicate meal. A decade ago these yards were crowded with platforms under construction, and the local roads and highways bustled with traffic bringing in the component parts and the people to assemble them. The rising price of oil initiated a rush offshore that transformed the local landscape and the local population. Today, there are few platforms under construction, and vacant buildings are abundant, as the offshore boom in the Gulf has played out with the impending exhaustion of the resource. Recreating the scenario in the coastal marsh, most of the major oil reservoirs offshore have been discovered, and the offshore platforms sited over them are busy sucking their viscous black libation and pumping it ashore. While the owners of the remaining resources continue to make profits, there is little call for new offshore construction or for the construction workers of a decade ago.

We turn northeast, away from the idle fabrication yards, to complete our circle and return to New Orleans. Out again we go, over the vast deltaic plain, interrupted only as we once more cross Bayou Lafourche and its linear settlements, until in the distance we see the meandering Mississippi. The river flows in an easterly direction toward New Orleans, and as we move in and parallel its tortuous progress, we pass enormous petrochemical plants and refineries attached malignantly to the riverbank, parasitically dependent on the lifeblood of the river to nourish and cool them and to carry away their waste. The massive cylindrical shapes connected by a confusion of exotically curved pipes, like the brass section of a giant band dropped from a great height, remind us that much of the oil harvested offshore and piped through lines under the linear canals that scar the wetlands ends up here being transformed into products like fertilizer, plastic, and those that power the very craft in which we fly. Called "Cancer Alley" by the local inhabitants, the plumes rising out of the abundant smokestacks fan out on the wind, obscuring our vision of the river and the plain, while the wastes discharged into the river remain invisible from our vantage point. In the distance the city lies outlined on the sweep of the plain. Then we are at the busy New Orleans airport, where we will land and later, on another aircraft, also powered by the end products of the system we have witnessed, depart the trembling prairie.

In our short trip we have seen the evolution of offshore oil and the land in which it occurred. The gradual growth of drilling into the marshlands, into the protected bays and estuaries, and then ever deeper offshore happened in the unique Louisiana deltaic plain. The millions of tons of steel for the offshore platforms and pipelines were fabricated primarily in this environment, and most of the billions of barrels of oil and trillions of cubic feet of gas that have come from the Gulf passed through it. Much of the oil is refined here, and the vast majority of support for this enormous network was, and is, situated in this fragile habitat.

This huge, hostile, fertile landscape was a major player in the evolution of offshore oil. Unlike most coastal regions of the world, the gradual transition from land to water allowed timid steps into the marine environment at a time when a full-fledged leap was technologically impossible. Likewise, the extended linear strips of available land, virtually all connected to water transportation routes, fostered the development of a network of small operations spread out over the deltaic plain, rather than the concentration of a few big fabrication and support facilities, as have occurred later, elsewhere in the world. But that is only a part of the picture. Evolving technology that not only made the push offshore possible but that created the demand for the petroleum exploited there must also be taken into consideration, as must the social changes and social movements that accompanied and, in many cases, led or opposed these technological trends. Likewise, the local, regional, national, and international economic and political climates, and changes in those climates, have a major role in this story.

It is the interaction of these factors over the developmental history of offshore oil in the Gulf and how the model of development nurtured in the Gulf failed in other coastal regions of the country that is the subject of this book. One consideration is essential to understanding the drive offshore. Before the enormous techno- logical and financial investments could be justified to move into the hostile environment of the marsh and the even more hostile environment of the open Gulf of Mexico, the dependence on and the demand for oil had to be enormous. There is no other way that the harvest of oil from under the open sea can be explained. Ac- cordingly, Chapter 1 (Developing the Demand) traces the dizzying spiral of the ascendancy of the importance of oil in American and world financial considerations. Starting with the first uses of oil and progressing to the point where offshore exploration and de- velopment became feasible, the chapter sets the stage for later events.

Chapter 2 (Early Technology and Politics of Offshore Oil) traces early offshore oil and gas activities as they made the transition from land to the marine environment, the emergence of a support sector that sustained the transition, and the political and economic forces that shaped that transition. Chapter 3 (Moving Offshore) follows the growth and maturity of the offshore industry as it emerged and developed technologically in the deltaic plain. Chap- ter 4 (Political Storm Clouds: OPEC, California, and the Embargo) lays out the signal events that affected offshore development and the major players in this drama. Chapter 5 (Boom and Bust in the Gulf) traces the rapid expansion of offshore activities in the Gulf of Mexico, particularly following the OPEC embargo of 1973–1974. Chapter 6 (Rising Political Controversy) examines the response to the increased push for offshore development brought on by the 1973 embargo and the emergence of forces that were eventually to shut down most of the Outer Continental Shelf in the United States to offshore development. And Chapter 7 (The Florida Con- flict) describes the opposition to offshore development in Florida and how and why that opposition has grown. The epilogue to the book sums up the current stalemate over offshore development and attempts to offer some useful observations and suggestions.

At a time when offshore oil development is welcome along some coastlines in the United States and is under Congressional and Presidential moratoria on others; when the process for balanc- ing the profits of oil against the effects of development on coastal communities and environments, which is a major part of the charge to the Department of Interior, has lost credibility in much of the nation; and when an ever-increasing dependency on petroleum energy, shrinking reserves, and a succession of ill-conceived quick fixes, premised on consuming our limited resources faster, is painting the country into a ideological energy corner, perhaps it is time to look back and try to understand this part of our past in order to plot our future.

Chapter 1

Developing the Demand

> An appraisal of the America of our times need not wait
> upon an historian of the twenty-first century to look
> back and note the primacy of oil. Oil has resettled our
> population, elected our presidents, swayed our foreign
> policy, and legislated our morality. Oil has not only
> fueled our wars—the two World Wars, Korea and
> Vietnam[1]—but is equally the sinew of our might in
> peacetime; it takes oil to be a superpower
> Solberg, Oil Power

Maybe so, but in order to understand the incredible invest ments of labor, technology, and capital associated with off- shore oil that will unfold in the pages of this volume, it is not enough to understand how we use oil and what it has done for us. We need to go deeper and examine how we came to be so depen- dent on oil and how such an enormous demand for the product was created. Oil was sold to an enthusiastic and willing public in the form of progress, freedom, and the American way. In the pro- cess individuals and corporations became wealthy beyond imagi- nation. The twentieth largest economic entity in the world today is Exxon, followed only by the gross national products of the nine- teen wealthiest countries in the world (Morgan 1986).

A part of the story must be understood against the backdrop of U.S. petroleum policy (or lack thereof) from the very beginning and the monopolistic structure of the oil industry from its inception, with the emergence of John D. Rockefeller and the cooperative policies of the international oil cartel between 1927 and 1973. The importance of the dates will become evident as the book unfolds. From this backdrop there are two basic trends that emerge and repeat themselves. First, without some type of control, which has almost never been effectively present, the history of the industry has been one of erratic booms and busts, as each new technological development in oil exploration (location of salt domes, seismic technology), production (drilling, transportation, refining) and consumption (kerosene, the automobile, etc.) spurred inequities between supply and demand.

Second, in the bumpy road to align supply and demand, virtually never has the finite nature of oil reserves been considered as a factor. Early during World War I, shortages of oil emerged, and the federal government worked with industry leaders to increase supply to meet demand. When the production of oil outstripped its consumption following World War I, instead of curtailing supply, the policy, in both the United States and the cartel, came to be to increase demand to meet supply, a feat that was accomplished completely only with the advent of World War II. Only in 1959 with the quota on foreign oil did the issue of curtailing supply emerge and then only with foreign oil and only to avoid a decline in the demand for the domestic oil. Then, with shortages following the oil embargo in 1973–1974, the focus became not curtailing demand but, with Nixon's Project Independence, increased production offshore and through the Trans-Alaska Pipeline; again, increasing supply to meet demand. Both strategies depend on increasing something a process that leads to a spiral of production and dependency. Only fleetingly have considerations been given to decreasing either supply or demand, and, for whatever reasons, they have been notably unsuccessful. Since both these trends have continued to the present and have directly affected offshore development, and since the uncontrolled and monopolistic nature of the playing field for oil persists, it is worth tracing their emergence.

EARLY HISTORY

Crude oil and pitch occur at natural seeps in many parts of the world and have been used in preindustrial times around the globe for a variety of purposes from fuel to road paving, from cosmetics to the caulking of boats. The first use of crude oil by European descendants in the United States was for medicinal purposes. "American Oil" or "Rock Oil," which was found in natural seeps, was touted as the cure for a variety of illnesses by the mid-1800s and was marketed nationwide. One of the more prolific production sites was found in Pittsburgh, Pennsylvania, and one of the imaginative entrepreneurs was found in the person of Samuel Kier. By 1848 Kier was bottling, selling, and extolling the "wonderful medical virtues" of his product, and by a decade later he had sold almost a quarter of a million bottles of it at a dollar each (Solberg 1976: 23). In 1854 a process for the distillation of kerosene from oil was patented by Abraham Gesner, a Canadian, and given that it represented a potential replacement for expensive whale oil, this additional commercial use for petroleum soon emerged in the American market. At the same time Kier was looking into other commercial uses for his product, and after experimenting with distilling his "Rock Oil," he built in 1854 what is generally believed to be the first commercial refinery in Pittsburgh. The refinery had a five-barrel capacity. By 1858 Kier and a local Pittsburgh firm were selling the refined product as an illuminant to New York City distributors. Demand for the product rose and soon petroleum was also finding a use in textile mills as a lubricant. Since supply was limited to that in natural seeps, the time was ripe for a technique that could provide more of the basic product (Giddens 1975).

Meanwhile in 1855, several speculators, after coming in contact with a bottle of "Rock Oil" and judging that it had commercial potential, had purchased land that had several crude oil seeps near Titusville, Pennsylvania. They soon formed the Pennsylvania Rock Oil Co. (the first petroleum company). After investigating the properties of their "Rock Oil," in the winter of 1957, the investors sent "Colonel" Edwin L. Drake, a vagrant with no technical or mining expertise, to Titusville to investigate the possibility of exploiting the seeps on their newly purchased land (Solberg 1976).

Drake arrived in Titusville in May of 1858 and, after unsuccessful attempts to dig a well at the site of one of the seeps, decided to adapt the local drilling technology that had been used to drill salt wells to petroleum. After several setbacks during 1858, Drake obtained the services of a local salt well driller in the spring of 1859 and, on August 27th of that year, struck oil. The first well produced an estimated eight to ten barrels a day, far more than was available from any other source. More importantly, Drake demonstrated that the expansion of the fledgling petroleum industry did not depend on natural sources of crude oil. In the ensuing decades the expansion of the market for petroleum products was literally to bring about the transformation of industrial society, and the search for oil touched off a process that would ultimately result in the world's most powerful multinational corporations. Drake himself was to die in obscurity in 1880. Even before then, however, two very important characteristics of oil and its use, which would be relevant many decades later in examining the effects of offshore development along the coast of the Gulf of Mexico, were to become apparent, one physical and one social.

CHARACTERISTICS OF OIL

One of the most important factors in the rise to prominence of oil and gas over earlier forms of energy lies in the physical characteristics of the substances themselves, specifically their fluid and gaseous makeup. This property is important for three reasons. First, fluids and gases can be more easily transformed through chemical processes (i.e., "refined" or "cracked") to produce a variety of finished products than can solids. While, for example, wood can be converted into charcoal and coal into coke, these are relatively limited transformations, and many of the volatile compounds associated with the original substances (potentially available for energy) are lost. Second, the ability to be moved in a fluidlike manner through pipes or tubes allows oil and its downstream products or gas to be easily injected into various types of energy and heat-producing devices at the downstream end of the energy use system (internal combustion engines, piston, jet, and turbine; stoves, furnaces, etc.). The alternative to shoveling coal into a firebox is to pump an alternative energy-producing substance there, and the internal combustion engine is obviously impossible without a combustible fluid or gas to inject into the cylinders. Finally, this same fluid property also allows efficient transportation of the product to market, either before or after refining, with a minimal amount of human handling and hence additional cost. Early oil wells were connected to collection points via pipe, and by the turn of the century the concept of moving oil by pipeline for at least short distances was common.

Today, oil is pumped from the ground into tankers for transportation to many areas of the globe. On arrival, oil is pumped from tankers into pipelines that carry it to refineries. From refineries the downstream products are pumped into pipelines, barges, or trucks for distribution to wholesale and retail outlets, where consumers increasingly pump it themselves into their own automobiles. All of this occurs without the "handling" (loading, unloading by human labor), packaging, and displaying associated with most products we buy.

Imagine an agricultural industry that harvested products in the field mechanically, pulverized and mixed them in central locations, and pumped the resulting product to stations where consumers would fill their own containers, bypassing the entire human harvest, initial packing, wholesale transportation, cooking/mixing, repackaging, displaying in grocery stores, and so on, and you can begin to get an idea of the advantage that oil offers as a product.

Furthermore, although oil and gas differ within limits in their chemical makeup, these differences can be accommodated by the refining process. So to use our agricultural analogy, suppose the process was limited to one crop to further simplify its production, processing, and distribution. It is precisely these properties of petroleum that have at least contributed to the enormous concentration of capital surrounding the harvest, processing, and distribution of petroleum and the vertical integration (control from the well to the automobile) of the multinational corporations that are the result of that capital accumulation.

THE SEARCH FOR OIL

It was the search for and production of petroleum that led to a distinctive social characteristic, which oil shares to a greater or lesser extent with other valuable resources, namely the cyclical nature of development. Only six years after the first oil well and only a few miles away, an instructive example of this, which was to be repeated many times, though perhaps not as strikingly, happened on the scene.

> On the flats at the bottom of this hill there was a tiny farm in January, 1865. A well was being drilled. On January 7, the well began to flow oil, a lot of oil! Men rushed here to drill wells of their own. The farm was sold, divided into lots, and Pithole City appeared by summertime. By September, as many as 15,000 people lived on this hillside. Over 50 hotels had been built, and stores, banks, offices and saloons filled the land half-way down the hill. By the next January, some of the wells had played out, and Pithole began to die. Fires burned much of the town, but people leaving quietly to go to richer oil fields led to the rapid decline of Pithole. After a few years, the land at the foot of this hill once again was only farmland along Pithole Creek. (Sign at the "Historic Museum" Pithole, Pennsylvania, 1992)

Pithole was the first oil boomtown. At its heyday, it was a thriving community, where fresh, raw, Chesapeake Bay oysters were served in local restaurants, and fortunes were being made. Pithole was the site of the first oil pipeline, six miles long connecting Pithole with the railroad junction at Miller Farm (Darrah 1972).

Pithole was also the site of the first oil bust. The new technology produced oil at a rate that outstripped the demand, and the price of oil throughout the early years of oil development was extremely unstable, at one point going from $20 a barrel to ten cents a barrel over the course of a year (Sampson 1975). But ultimately, it was the local availability, not the price of oil, that signaled Pithole's demise. During the fall of 1865 Pithole's wells produced over 900,000 barrels of oil, about a third of the annual U.S. production for 1865. By the following spring production had fallen and by the fall of 1866, virtually ceased. The reservoir had been drained through overproduction, and the population soon followed the fate of the oil.

The growth and decline of the community over its 500-day life cycle illustrates several important aspects of the extraction of natural resources that have relevance for understanding offshore development as it began to unfold in the Gulf of Mexico seven decades later. One of the primary characteristics of resource extraction is that extractive activities must locate where the resource is (Bunker 1984; Gramling and Freudenburg 1990). Since there is no necessary correspondence between where natural resources like oil occur and where human populations have settled, extractive activities, unlike manufacturing enterprises, frequently cannot take advantages of existing infrastructure and concentrations of support sectors of the economy. In the absence of this shared location advantage, extractive undertakings must provide everything necessary to support the primary activity and as a result must frequently rebuild the local physical and social environment. The construction of facilities necessary to house and provide amenities to a population of 15,000, as well as to drill for and transport an ultimate total of 3.5 million barrels of oil in an isolated rural setting in the Pennsylvania in 1865, make Pithole an excellent example.

Pithole and the surrounding early development in northern Pennsylvania also illustrate several other attributes of extractive activities. Not only are resources like oil notorious for price fluctuations because of their linkage to the commodity markets in which they are bought, sold, and traded, but ultimately they become exhausted. As this happens in local environs, because the infrastructure built to support one specific extractive activity may not be useful for other activities, communities, facilities, and the human capital of skills, investments, and networks of supply become obsolete and, to greater or lesser extent, may disintegrate. In the case of Pithole, disintegration was total. In addition to the characteristics of oil and the search for oil, the monopolistic structure of the oil industry was also to emerge early and persist through time.

ROCKEFELLER, STANDARD OIL, AND EARLY MONOPOLY

In spite of the uncertainties associated with early oil development, not everyone was cautious about entering the fray. With boundless energy, the mind set of an accountant, and a ruthless determination, John D. Rockefeller leveraged a small refinery in Cleveland, in which he originally owned a partnership, into a monopoly, the remains of which continue to dominate the world economy today. Rockefeller worked long hours, apparently had great organizational ability, and, when possible, operated in total secrecy, but the key to his cornering the petroleum market in the late 1800s was his manipulation of the railroads.

Following the Civil War, the railroads emerged as a primary industrial power in America. Collective owners of hundreds of millions of acres of land, not only did the railroads bring enormous personal fortunes to their owners (e.g., Vanderbilt) and have the ability to control the fortunes of regional enterprises (Cochran and Miller 1942), but, as Cronon (1991) demonstrated, because of the (previously unavailable) mobility they allowed, it was possible for centers of specialization and transition to emerge. Chicago, as a result of the railroads, emerged as both the center and transition point for the exploitation of the tall grass prairie, more euphemistically called the "development of the great west." Although Pittsburgh was the site of the first oil refinery, Cleveland emerged as the early center of refining in the United States.

By 1867 Rockefeller had swallowed up twenty-two of his Cleveland competitors and, as a result, controlled half of the refinery potential in the world. Soon, the newly formed Standard Oil Company was negotiating rebates on oil shipped east with both railroads that served Cleveland (Solberg 1976). The rebates allowed Standard to ship oil at a competitive advantage, and the profits went into buying out the competition. Using the same strategies, Standard quickly moved outside of Ohio, and by 1882, the Standard Oil Company had become the Standard Oil Trust, an interlocking network of over thirty companies, which by the late 1880s was operating in all states and moving aggressively into the international market (Solberg 1976).

Upon encountering competition from a marketing firm formed by two of Alfred Nobel's brothers and financed by the Rothschild

Paris banking interests, Rockefeller created his own international firms, the first, the Anglo-American Petroleum Company in 1888. Before long, however, realizing that competition was counterproductive, Standard, the Nobels, and the Rothschilds entered into deals that divided up the world market in patterns that continue to the present. This international cooperation set a precedent that was to have important implications throughout the twentieth century and that illustrated yet another characteristic of the development of oil. Where monopoly was impossible, cartels were the answer. These monopolistic corporate cooperations controlled the world oil market, with varying degrees of success, for over eight decades, until bowing to a new type of cartel based on oil-producing nations.

Another of Standard's strategies, developing downstream linkages for the marketing of oil, would later turn into full vertical integration, which became a distinguishing characteristic of the industry. Eventually, multinational oil companies controlled their product from the initial leasing, through drilling, production, refining, transportation, wholesaling, and retailing.

Throughout this time period Standard was acquiring interests, often secretly, in key regional wholesale corporations and marketing Standard products. Although petroleum was primarily used at this time in the form of kerosene, by the turn of the century, as the first automobiles appeared on the scene, Standard's marketing network was already nationwide (Solberg 1976). Rockefeller who had started in refining and moved into marketing, strangely, never took the additional step toward upstream linkages—petroleum production.

In response to the public outcry against the practices of Standard and some of the other major trusts (primarily the railroads), Congress enacted the Interstate Commerce Law in 1897 and several years later the Sherman Antitrust Law, neither of which were enforced by then President Cleveland or his successor McKinley, who received $250,000 in campaign contributions from Standard Oil (Flynn 1932).

In 1901 McKinley was assassinated, and Teddy Roosevelt entered the White House. Reading the winds of change, Roosevelt directed the Justice Department to prosecute both Standard Oil and the railroads under the Sherman Act. Roosevelt and his successor Taft were both out of office by the time the legal battle was fought all the way through the Supreme Court. It was in 1911, the same year that gasoline sales first topped those of kerosene and a year before the Model T was to go into production, that the Supreme Court ordered the Standard Oil Trust broken up into over twenty companies (Solberg 1976).

SPINDLETOP AND MOVEMENT TO THE GULF

While the Standard Oil Trust case was being fought through the courts, events were unfolding along the Gulf of Mexico that were to affect Rockefeller's attempt to control the U.S. petroleum market and that were to have major implications for the movement offshore into the Gulf. The battleground was Texas, and James Henry Hogg was the major protagonist. Hogg, first elected to the office of attorney general and then in 1890 to governor, was a self-proclaimed opponent of the big trusts that, in the wake of reconstruction, were monopolizing commerce in Texas.

The big trusts boiled down to the railroads and Standard Oil, and Hogg's basic tools were a tough state antitrust law, which effectively made it impossible for Standard to operate legally in the state, and the Texas Railroad Commission, which was given broad regulatory powers. At Hogg's insistence the commission was given such wide powers that not only could it regulate railroads but also later take over the regulation of the state's oil, although no one had any reason to believe that Texas had any oil at that time. In fact, John Archbold, one of the directors of Standard, later to become president when Rockefeller stepped down, is said to have laughed at the idea and offered to drink all the oil found west of the Mississippi (Solberg 1976). While the antitrust law was primarily aimed at Standard's distribution network and close cooperation with the railroads, its latent effect was to hobble Standard in its primary activity of refining since Texas did prove to have oil.

The world's first major petroleum reservoir, Spindletop, was tapped on January 10, 1901, near Beaumont, Texas. Named for the hill that lay over it, on which spindly pines grew,[2] Spindletop transformed the petroleum industry in the United States. The state's antitrust laws forced Standard Oil to operate in secrecy in Texas, and new players, including Andrew Mellon from Pittsburgh and former governor Jim Hogg, moved into the action. Because the impetus for their entry into the industry began with the discovery of a vast petroleum reserve, these new players did not share Rockefeller's reluctance to become directly involved in the production of oil. Furthermore, in order to compete with Standard's monopoly of downstream products, they were forced to enter all phases of petroleum production, refining, and distribution, in order to market their products. As a result the corporations emerging in Texas, Gulf and the Texas Company (later Texaco), did so as fully vertically integrated entities, controlling their product from the well to the retailer. As Nicholson noted forty years later:

> Until discovery of the vast field [Spindletop], big names in the industry were scarce. Yet this huge field marked the beginning of such companies as The Texas Company and Gulf Oil Corporation, and furnished the nucleus for Magnolia Petroleum Company [the first operator on the Outer Continental Shelf]. (Nicholson 1941a, p. 17)

Spindletop and subsequent Gulf Coast discoveries moved the focus of the petroleum industry out of the eastern United States and squarely into Texas and the coastal Gulf of Mexico. Wildcatting operations came into their own with Spindletop, and many Texans, including Howard Hughes and H. L. Hunt, got their start there (Solberg 1976). The unmistakable rough and tumble cowboy image, which was stamped on the emerging "culture" of the industry (Freudenburg and Gramling 1994a) during its development in Texas,[3] persists to the present. Throughout the next several decades wildcatting continued successfully in Texas, many of the strikes accompanied by Pithole-like booms, and, like boomtowns everywhere, a raw, reckless way of life flourished. Through it all there emerged a new industrial class of men, those who had struck it big, had newfound wealth and power, and little experience with either. The Texas oil millionaire, long on arrogance and conspicuous consumption and short on manners and knowledge, became almost a caricature outside of the oil patch and concurrently a hero to the emerging industrial subculture.

WORLD WAR, DEPLETION, AND THE INTERNATIONAL CARTEL

With the push into Texas and with oil production moving away from the small independent producer that had been common on the East Coast, some of the more powerful of the emerging producers began to cooperate for their own political interests. The independents had long argued that, because they "produced" a finite commodity, using up their assets and, hence, capital as they went, they should be allowed to account for this when computing taxes. In 1913 Congress allowed such a depletion, allowing a 5 percent deduction of the gross value of production for mining and oil interests. This was changed in 1916 to the disappointment of the producers, limiting it to a "reasonable allowance" based on the cost of actual exploration. By this time, however, several factors had improved the position of the producers. First, many of the big players had become vertically integrated companies, some to the extent that they controlled petroleum from exploration to the gas tank. Standard Oil of New Jersey, the original linchpin of the Standard Oil Trust, had been taken in this direction by its new president, A. C. Bedford, as had other past members of the trust and most of the big operators in Texas (Gulf, Texaco). Thus, these powerful interests now had a stake in the depletion allowance.

Second, the United States entered World War I in 1917, and the War Industries Board with its National Petroleum War Service Committee and its director Mark Requa actively encouraged the cooperation of the major players in the oil industry in order to be able to supply the war effort in Europe. As Solberg put it:

> Overnight, the industry, with President Wilson's support, was doing what the Sherman Antitrust Act forbade. Six years after the dissolution of Standard Oil's trust, its chief executive was in Washington helping direct industry's cooperation with government. (Solberg 1976: 73)

The end results of this cooperation was that the Revenue Act of 1918 allowed oil producers to base the depletion on a "reasonable allowance" of either the cost of discovery or of the fair market value of the discovery, which for most companies averaged between 28 and 31 percent of their gross income (Solberg 1976).[4] This enormous tax break, initiated during the last year of the war, greatly encouraged exploration and drilling, as it was intended to. After the war, however, following initial confusion with the abrupt dismantling of wartime controls by the Wilson administration and periodic shortages for several years (Nash 1968), the increased levels of production of foreign[5] and Gulf Coast oil quickly led to a glut on the world oil market. In addition to the depletion allowance, closer cooperation, at the urging of Requa, led Rockefeller and others to the formation of the American Petroleum Institute immediately after the war in 1919.

Oil, by this time, was truly an international activity. By 1920 world annual production of crude oil was up from 6,000 barrels in 1859 (the year Drake drilled the first well) to 689 million barrels (*Oil Weekly* staff 1946a), or approximately the current annual capacity of the Trans-Alaska Pipeline. The supremacy of Standard had not only been challenged in the United States (by Gulf and the Texas Co.), necessitating a sharing of the power, but several international corporations had risen to dominance. Chief of these was the Anglo-Dutch Shell Oil Company, led by Sir Henry Deterding. Not only had Shell moved into the West Coast of the United States, but it had massive holdings and markets worldwide. By the mid-1920s Shell was the world's largest producer. Production outstripped demand, and serious international competition, including a price war in India, led some of the major players to realize that competition was not in their best interests. In September of 1927, Deterding invited the chairman of Standard of New Jersey, Walter Teagle, and the chairman of Anglo Persian, the other major British company, to a castle he had leased at Achnacarry, Scotland. They were soon joined by representatives of other major companies. The result of these meetings was the Pact of Achnacarry or the "as is" agreement, as it came to be known (Solberg 1976; Engler 1961; Ghanem 1986). This illegal (at least under United States laws) agreement in effect created a world oil cartel, where the current division of the world market was affirmed, a number of production quotas were set, and the necessity for competition was eliminated. Each market could be supplied from the nearest source, but a constant price was assured. The cost, no matter where supplied from or to, would be based on the current cost of the oil in the Gulf of Mexico (read Texas) market, plus the cost to ship from the Gulf.[6] The agreement ensured orderly, carefully controlled marketing and development, and though uncovered by the Federal Trade Commission in 1952 (Federal Trade Commission 1952), the principles, for all practical purposes were in effect until the oil embargo of 1973–1974, though it had begun to unravel after the formation of the Organization of Petroleum Exporting Countries (OPEC) in 1960.

THE DEPRESSION AND PRORATIONING

While the Pact of Achnacarry settled the problems of world competition, the problem of overproduction was not finished. A little over two years later and shortly after newly elected President Hoover withdrew federal lands from leasing in an attempt to control overproduction (Nash 1968), the United States stock market crashed, securities lost $26 billion in value, and the emerging world economy went into a depression. For the first time the petroleum industry had to seriously consider supply-side regulation, and the word that came down was prorationing. The idea behind *prorationing* was that all producers would cut their production and thus *ration* their output but would retain their *proportion* of the total output. The smaller producers with less of a profit margin were opposed to the idea, while the major producers were in favor of it in order to cut supplies and raise prices.

A year after "Black Friday" the supply situation got totally out of hand with the discovery of the largest oil field to date in east Texas. The field was too big for the majors to control, and oil began to flood into the market. The price of oil fell first to ten cents a barrel and then to two cents a barrel. The Texas Railroad Commission tried to control the flow, but producers started transporting their crude to other states. As Solberg describes it:

> As days passed, more and more such "hot oil" poured onto the markets. Down went prices until water cost more than oil—two cents a barrel for crude oil, four cents a gallon for

"Eastex" gasoline distilled at the field's ninety-odd small refineries and offered at pumps with a free chicken dinner, a
dozen eggs, or a crate of tomatoes with a fill-up. When officials tried to shut down small producers and stop trucks, riot-
ing erupted. Governor Ross Sterling proclaimed martial law.[7]
(Solberg 1976: 125)

Franklin Roosevelt took office in 1933, inheriting the growing
problems from Hoover. Although a number of antitrust actions
against the larger firms in the oil industry were pursued early in
Roosevelt's administration and although Roosevelt's Secretary of
Interior, Harold Ickes,[8] was convinced the oil industry should be
regulated like public utilities, the federal government's role was
minimal. The attempted resolution of the problem came in the
form of the Connally Hot Oil Act, which was supported by the
then House majority leader, Sam Rayburn of Texas. The act forbid
the interstate sale of oil and allowed the states involved to attempt
to control production through prorationing at the state level. This
was in marked contrast to the way in which most industries were
handled under the National Recovery Administration (declared
unconstitutional by the Supreme Court in 1935) under Roosevelt.
Neither approach appeared to be particularly successful, for both
the New Deal and the oil industry. It took war to achieve their
goals (Solberg 1976).

If attempts to limit supply were generally not very productive,
efforts to increase demand met with more success. Before and during World War II corporations controlled primarily by General
Motors, but also with interests by several of the major oil companies and rubber companies, systematically bought up and
scrapped mass transit systems in major cities throughout the
country (Snell 1974). The replacement for these systems was, of
course, buses built by General Motors, and what was quickly becoming the quintessential American artifact, the automobile.

MOVEMENT INTO THE MIDDLE EAST

During the same time period that the major oil companies were
entering into agreements to control the world oil market and

scrambling to regulate oversupply and build demand, discoveries were emerging in the Middle East that almost five decades later were to prove to be the undoing of that control. Oil was discovered in what is now Iraq in 1927, in Bahrain in 1931, and in Saudi Arabia and Kuwait shortly thereafter. Given the new spirit of cooperation produced by the Achnacarry agreement and with the support of the U.S. government, five of the major U.S. oil companies entered into a consortium with British, Dutch, and French interests, forming the Turkish Petroleum Company. The company, later renamed the Iraq Petroleum Company, was formed to exploit the reserves of the former Ottoman Empire, which had been dismembered after World War I. The consortium and a series of additional agreements between the original companies and new entrants into the Middle East soon led to a complex series of agreements and interlocking directories that virtually completely controlled oil production in the Middle East (Kaufman 1978).[9] Although these finds were later to prove to contain the majority of the world's proven crude oil reserves and to become the focus of much foreign policy by the industrialized nations of the world and the source of a war in 1990–1991, their full development would have to wait until another war, World War II, was over.

WORLD WAR II

With the breakout of war in Europe in 1939, the oil crisis again moved quickly from oversupply to the need for increased production. Roosevelt put Ickes in charge of the new thrust delegating him Petroleum Coordinator for National Defense. Ickes convinced (with some pressure from Roosevelt) the Department of Justice to suspend antitrust suits against the oil industry, came out against oil exports to Japan (which continued through 1940 and the first part of 1941), and, because of the increasing danger to tankers in the Gulf from German submarines, pushed for pipelines from the midcontinent oil fields to supply the East Coast. Exports to Japan were halted in 1941, leading directly, some believe, to Pearl Harbor (Solberg 1976). By 1943 the "Big Inch" pipeline (24 inches in diameter) was supplying oil to the East Coast from the fields in Texas and Oklahoma, averting a shortage and coincidentally transforming the long-range transportation of oil.

World War II was fought with oil and over oil. Japanese strategy throughout hinged on imports for their island nation, particularly to supply the navy, and allied strategy, on denying those imports. Rommel's push into Africa, threatening Middle Eastern supplies led to Allied response and, once the threat was over, for Ickes (now "Petroleum Administrator for War") to push for a pipeline from the Saudi fields to the Mediterranean and for greater United States access to Middle Eastern oil (Nash 1968). This latter thrust, which eventually led to greater dependence on this source, was to have major implications for offshore development three decades later when the source was interrupted. The war period also saw the first major exploration of the Alaskan arctic. Ironically, a map published in 1946 in *The Oil Weekly* (Thomas 1946) for a proposed pipeline route for then still-undiscovered oil near Barrow, Alaska, closely parallels the current Trans-Alaska Pipeline. The Trans-Alaska Pipeline was to become coupled with the controversy over increased offshore exploration and production three decades later.

The war saw the organization of oil production under Ickes and enormous increases in actual production. Between 1935 and 1948, United States oil production doubled (see Table 1.1). Following the war, three fundamental policy initiatives and a cluster of incentives favoring the growth of the automobile as the primary form of transportation ensured that consumption would parallel this trend.

First, in addition to the push away from public transportation caused by the cooperative dismantling of systems across the country by General Motors and some of the major oil and rubber companies (Snell 1974), Solberg notes the variety of factors that pulled consumers to the automobile.

> First of all, an automobile owner was required to pay very small taxes on it. Second, the gasoline tax, while higher, went directly to build more roads used by the private car. Third, both federal and state governments subsidized the highway construction program while abandoning support of public transportation. Fourth, the social and environmental costs of automotive growth-highway deaths, traffic snarls, smog,

Table 1.1. U.S. and World Petroleum Production (Thousands of Barrels)

YEAR	U.S.	WORLD	U.S.%	YEAR	U.S.	WORLD	U.S.%
1860	500	509	98.23	1900	63,621	149,137	42.66
1861	2,114	2,131	99.20	1901	69,389	167,440	41.44
1862	3,057	3,092	98.87	1902	88,767	181,809	48.82
1863	2,611	2,763	94.50	1903	100,461	194,879	51.55
1864	2,116	2,304	91.84	1904	117,760	217,948	54.03
1865	2,498	2,716	91.97	1905	135,602	215,091	63.04
1866	3,598	3,899	92.28	1906	127,565	213,263	59.82
1867	3,347	3,709	90.24	1907	167,889	263,957	63.60
1868	3,646	3,990	91.38	1908	182,988	285,257	64.15
1869	4,215	4,696	89.76	1909	183,171	298,709	61.32
1870	5,261	5,799	90.72	1910	209,557	327,763	63.94
1871	5,205	5,730	90.84	1911	220,449	344,361	64.02
1872	6,293	6,877	91.51	1912	222,935	352,443	63.25
1873	9,894	10,838	91.29	1913	248,446	385,345	64.47
1874	10,927	11,933	91.57	1914	265,763	407,544	65.21
1875	8,788	9,977	88.08	1915	281,104	432,033	65.07
1876	9,133	11,051	82.64	1916	300,767	457,500	65.74
1877	13,350	15,754	84.74	1917	335,316	502,891	66.68
1878	15,397	18,417	83.60	1918	355,928	503,515	70.69
1879	19,914	23,601	84.38	1919	378,367	555,875	68.07
1880	26,286	30,018	87.57	1920	442,929	688,884	64.30
1881	27,622	31,993	86.34	1921	472,183	766,002	61.64
1882	30,350	35,704	85.00	1922	557,531	858,898	64.91
1883	23,450	30,255	77.51	1923	732,407	1,015,736	72.11
1884	24,218	35,969	67.33	1924	713,940	1,014,318	70.39
1885	21,859	36,765	59.46	1925	763,734	1,068,933	71.45
1886	28,065	47,243	59.41	1926	770,874	1,096,823	70.28
1887	28,283	47,807	59.16	1927	901,129	1,262,582	71.37
1888	27,621	52,165	52.95	1928	901,474	1,324,774	68.05
1889	35,164	61,507	57.17	1929	1,007,323	1,485,867	67.79
1890	45,824	76,633	59.80	1930	898,011	1,410,037	63.69
1891	54,293	91,100	59.60	1931	851,081	1,373,532	61.96
1892	50,515	88,739	56.93	1932	785,159	1,309,677	59.95
1893	48,431	92,038	52.62	1933	905,656	1,442,146	62.80
1894	49,344	89,337	55.23	1934	908,065	1,522,218	59.65
1895	52,892	103,692	51.01	1935	996,596	1,654,488	60.24
1896	60,960	114,199	53.38	1936	1,099,687	1,791,540	61.38
1897	60,476	121,993	49.57	1937	1,279,160	2,039,231	62.73
1898	56,122	124,979	44.91	1938	1,214,355	1,988,041	61.08
1899	57,071	131,147	43.52	1939	1,264,926	2,085,444	60.65

Table 1.1. continued

YEAR	U.S.	WORLD	U.S.%	YEAR	U.S.	WORLD	U.S.%
1940	1,353,214	2,142,189	63.17	1966	3,027,763	12,021,786	25.19
1941	1,402,228	2,247,549	62.39	1967	3,215,742	12,914,340	24.90
1942	1,368,645	2,060,353	66.43	1968	3,329,042	14,146,318	23.53
1943	1,503,614	2,300,579	65.36	1969	3,371,751	15,222,511	22.15
1944	1,677,753	2,621,934	63.99	1970	3,517,450	16,718,708	21.04
1945	1,710,275	2,768,885	61.77	1971	3,453,914	17,662,793	19.55
1946	1,764,561	2,849,953	61.92	1972	3,455,368	18,600,745	18.58
1947	1,856,987	3,022,139	61.45	1973	3,360,903	20,367,981	16.50
1948	2,020,185	3,443,234	58.67	1974	3,202,585	20,537,727	15.59
1949	1,841,940	3,404,142	54.11	1975	3,056,779	19,502,335	15.67
1950	1,973,574	3,803,027	50.95	1976	2,976,180	21,191,540	14.04
1951	2,247,711	4,282,730	52.48	1977	2,985,360	21,900,695	13.63
1952	2,289,836	4,531,114	50.54	1978	3,178,216	22,158,251	14.34
1953	2,357,082	4,798,055	49.13	1979	3,121,310	22,765,050	13.71
1954	2,314,988	5,016,591	46.15	1980	3,146,365	21,746,164	14.47
1955	2,484,428	5,625,659	44.16	1981	3,128,624	20,380,505	15.35
1956	2,617,283	6,124,676	42.73	1982	3,156,715	19,375,125	16.29
1957	2,616,901	6,438,444	40.64	1983	3,170,999	19,210,924	16.51
1958	2,448,987	6,607,750	37.06	1984	3,249,696	19,753,368	16.45
1959	2,574,590	7,133,238	36.09	1985	3,274,553	19,488,948	16.80
1960	2,574,993	7,674,460	33.55	1986	3,168,252	20,329,092	15.58
1961	2,621,758	8,186,213	32.03	1987	3,047,378	20,169,163	15.11
1962	2,676,189	8,881,858	30.13	1988	2,979,123	21,045,249	14.16
1963	2,752,723	9,538,346	28.86	1989	2,778,773	21,383,918	12.99
1964	2,786,822	10,309,644	27.03	1990	2,684,687	21,664,210	12.39
1965	2,848,514	11,062,515	25.75	1991	2,707,039	21,464,024	12.61

Source: *Oil Weekly* staff, 1946; American Petroleum Institute, 1986, 1993.

high-speed roads through parks and residential neighborhoods, for instance—were accepted by the community at large. Finally and most important, all political and economic institutions, from the Congress in Washington to the mortgage bank down the street, supported an automobile-dominated organization of urban and suburban development. In short, just about all the resources in the society stood committed to producing automobility, no matter what the consequences. (Solberg 1976: 142)

More cars, bigger cars, bigger engines, and more gasoline consumption were the story through the 1950s and 1960s (see Table 1.1) and were major factors in the growing consumption of crude oil.

This greater freedom in transportation led to the flight from the central cities to the suburbs, a trend reinforced by federal policy. The Federal Housing Administration and the Veterans Administration (under the terms of the G.I. bill) made low-cost housing available, and both the laws and implementation of guidelines favored the new suburbs over the central cities (Solberg 1976). Not only did commuting to work from the suburbs consume petroleum products, but the new single-family dwellings heated by fuel oil added comfort at the expense of increased energy consumption.

A second basic policy consideration, which added fuel to the consumption fire, was the postwar policy for the rebuilding of Europe and Japan. The Marshall Plan for the rebuilding of Europe called for a switch from coal, the dominant source of power before and during the war, to oil. Given the massive increase in consumption of oil this plan would lead to and given the European interest in the Iraq Petroleum Company, the strategy became to supply much of this oil from the Middle East. Under occupation, Japan was also encouraged to switch its industries from coal to oil, and much of this was also supplied from the Middle East through U.S. oil companies (Engler 1977). Thus, the postwar recovery of Europe and the miraculous development of Japan were fueled by oil.

A third major force for increased consumption was Eisenhower's Interstate Highway and Defense System. Justified during the growing cold war as a necessity for rapid evacuation of the cities in case of nuclear attack, the $27 billion system not only moved transportation of goods away from more energy-efficient systems (railroads and waterways) and onto the highways but exacerbated the flight to the suburbs. Over half of the $27 billion was actually spent in the cities, where increased access and egress to the central cities through urban freeways, created with little or no social planning, brought new traffic problems, destroyed neighborhoods, displaced populations, and further reduced the appeal of the city as a place to live, adding new impetus to the flight to the suburbs (Solberg 1976; Llewellen 1981).

In addition to these factors affecting the increased consumption of petroleum, the aftermath of World War II also brought the "Tidelands" disputes, a major political battle between the federal government and the coastal states over the ownership of offshore lands. It was the Tidelands dispute that saw Ickes resignation under the Truman administration. The incident was Truman's nomination of Edwin Pauley to be Undersecretary of the Navy early in 1946. Ickes contended that Pauley had worked vigorously on behalf of the states and had tried to bribe him (Ickes) to drop a pending federal suit against several coastal states (including Louisiana, California, and Texas) to secure federal ownership of offshore lands (Solberg 1976; Nash 1968). In the ensuing conflict with Truman, Ickes resigned, and one of the most determined champions of the stewardship of federal lands retired from the scene. Two pieces of legislation arising out of these disputes, the Submerged Lands Act and the Outer Continental Shelf Lands Act, both passed in 1953, were to become the underpinning of the movement offshore in search of oil (chapter 2).

Thus, by the mid-1950s a dizzying spiral of petroleum consumption had been put in motion. The economic consequences of this were not to become apparent for two decades, and realization of the environmental consequences took even longer. As Catton (1986) has noted, the growth of technology made possible the exponential increase in the amount of energy the average American could consume, reinforcing the supply-demand cycle noted earlier. This increased consumption potential led to the search for oil in the most unlikely of places, under the open seas. The evolution of the technology and the legal structure to allow the recovery of oil under the open seas are covered in chapter 2.

Chapter 2

The Early Technology and Politics of Offshore Oil

The first over-water drilling for oil occurred in 1896 at Summerland, California, from a wooden pier into the Pacific Ocean. The success of this endeavor brought quick emulation, and by 1902 a photograph of the Summerland development "at its peak" shows over forty derricks on piers over the Pacific (Lankford 1971). Although technically the drilling was over water, little modification of existing land-based technology was required because of the direct land connection, and over-water technology moved slowly in the Pacific.

It was the first major oil reservoir discovered, at Spindletop in Texas, that not only transformed the petroleum industry (chapter 1) but also focused attention on the Gulf Coast and on the adjacent state of Louisiana, where most of the evolution in offshore technology was to take place. This technological adaptation to the marine environment progressed over a decade later with the development of the Caddo Lake field in northwestern Louisiana. By 1905 successful oil wells had been drilled surrounding the lake, and in 1910 the Board of Commissioners of the Caddo Levee District advertised for bids on leases on the bed of the lake. The big bidder was the newly formed Gulf Oil Company, which leased 8,000 acres, and a decade and a half later had a sizable fleet in operation on the lake (Forbes 1946; Lankford 1971).

The Caddo Lake development required the solution of several technical problems, which were later to have implications for the move offshore in the Gulf of Mexico. First, the downhole pressures in the reservoirs were beyond anything encountered before, and many of the early wells blew out, caught fire, and burned uncontrollably. The problem became so bad that it resulted in several state laws regulating drilling procedures and a withdrawal of further public lands in the area for development by the U.S. Department of Interior's General Land Office (Lankford 1971). The movement into the Caddo field was quickly followed by the emergence of increasingly reliable blowout preventors (Brantly 1971), a necessary precondition for movement into the high-pressure reservoirs along much of the Gulf Coast.

A second requirement of the Caddo development was the ability to drill and produce without direct connection to land, although the initial exploration and development was again little more than creative utilization of existing land-based drilling technology. Platforms were constructed on pilings driven into the lake bottom, and drilling equipment was barged to the site. Finally, another technological innovation, underwater pipelines connecting the production wells to one of four gathering stations on the lake was a precursor of the collection lines currently used offshore. The Summerland and Caddo experiences were the start of the offshore technology.

INTO THE MARINE ENVIRONMENT

This technology moved ahead with the development of the Maracaibo field, one of the largest ever discovered in the Western Hemisphere, ultimately producing approximately 4.6 billion barrels of oil (Lankford 1971). Lake Maracaibo in northern Venezuela differed from Caddo Lake in three important ways. First, it was larger, approximately two thirds the size of Lake Erie. Second, it was deeper, up to 120 feet. Third, it was connected to salt water. It was this last characteristic that required the first modification in drilling practices, when in 1924 Lago Petroleum Company[1] brought in the first well in Lake Maracaibo. The basic problem faced was the teredo, or shipworm, which thrived in the brackish water of the lake and which destroyed wooden pilings in six to eight months. After experimenting with a variety of local timber to no avail, the material of choice became concrete. Concrete pilings were driven in as legs and as support under the drilling machinery, and a concrete base was cast in place (Lankford 1971). Eventually, the concrete base was replaced with a steel deck, which eliminated the need for the central support pilings, and, as greater water depth required, the pilings were replaced with hollow steel reinforced concrete caissons, a technology that has further evolved in the North Sea and North Atlantic into gravity-based platforms.

Finally, the number of over-water wells used to tap the huge reservoir under Lake Maracaibo, each of which required the setup of the drilling machinery on a platform and later breakdown of that same machinery, led to the steam drilling barge. During drilling the derrick continued to be mounted on the platform, but the power supply and the machinery for drilling was mounted on a barge, which could be easily moved from platform to platform. This increased mobility led to considerable savings and provided a model for the drilling tenders first used in the shallow waters of the Gulf of Mexico.

EARLY DEVELOPMENT ALONG THE GULF COAST

Following the massive discoveries in east Texas and as nondestructive exploratory (seismic) techniques evolved in the 1920s, numerous salt domes were located by geophysical exploration along the Louisiana and Texas Gulf Coast. Access to these discoveries was a problem, however, since as Lankford (1971: 1377) noted, "these inviting prospects were often situated in very uninviting terrain—marshes, swamps, and the shallow open waters of bays and lakes." Neither the traditional land-based drilling techniques nor those evolving in Caddo Lake or Lake Maracaibo worked very well. The problem was the extremely fine silt laid down in great depth over time by the Mississippi River in its various configurations. The lower portion of the State of Louisiana (literally the deltaic plain) was comprised primarily of this material. Over time decayed vegetation also became a major element of this "soil." Since the decayed vegetation absorbs water and the silt acts as a lubricant, the resulting "highly aqueous organic ooze" (Russell 1942: 78) has a consistency slightly stiffer than chocolate pudding. When found as shallow marsh in the early days of coastal exploration it did not provide the support that traditional land-based drilling activities required, and elaborate preparation of the drill site was necessary. When found underlying shallow open water (as was the case in much of coastal Louisiana), the bottom was even more problematic. Since the silt layer can be very thick, pilings driven into it have no solid foundation, relying entirely on the friction between the pilings and the mucky bottom. As Lankford (1971) notes, since platforms built on such pilings cannot be cross-braced below the water line and since drilling machinery produces significant vibrations, up to 250 pilings were needed for a single exploratory platform. As interest grew in deeper prospects, these techniques became cost inefficient for exploration, since the nonrecoverable cost of platform construction was a constant regardless of the outcome of the drilling. Nevertheless, exploration and development did go forth along the Texas coast and the more inhospitable Louisiana coast, in spite of the nature of the terrain.

The technological breakthrough that facilitated exploration in the shallow coastal waters and marsh came in 1933 with the Texas Company's introduction of the submersible drilling barge. The Texas Company, like Gulf, had been born out of the discoveries at Spindletop and protected from the Standard Oil Trust, and later Standard of New Jersey, by the tough antitrust laws in Texas. Quickly forced to become a vertically integrated company to compete with Standard but lacking resources to compete fully with Standard on the world market, the company initially saw its interests and greatest potential as lying along the Gulf Coast. As such, Texas was one of the firms most interested in the new potential in the environs of south Louisiana and in the late 1920s leased large tracts of marsh from the Louisiana Land and Exploration Company, as well as submerged lands under the contiguous lakes and bayous from the state of Louisiana (Lankford 1971). The problems with exploration and development remained, however, and Texas began to explore other solutions. During a routine patent search after a Texas employee submitted plans for a mobile drilling barge in 1932, it was discovered that Louis Giliasso had been issued a patent for a submersible drilling barge four years earlier, in 1928. Giliasso was eventually located the following year, operating a bar in Panama, and an agreement for the use of the patent was struck.

In theory, the submersible barge could be towed to the drilling site and sunk. Since the water at these locations was shallow, the barge, while resting on the bottom, provided a stable base for drilling. The elevated derrick was positioned over a rectangular vertical slot extending from one end to the center of the barge. Drilling proceeded down through this slot. Once drilling was complete, the barge could be raised and moved to a new location. The Giliasso (as the first barge was appropriately named) was built in Pennsylvania and brought down the Ohio and Mississippi rivers to Lake Pelto, Louisiana, one of the protected estuaries behind the barrier islands along the coast of Louisiana.[2] There on a state lease on November 17, 1933, the Lake Pelto No. 10 was spudded. The barge was a success, and a new era in marine drilling had begun (Williams 1934). Although in retrospect the principle seems simple, the submersible drilling barge revolutionized the movement into the marine environment. Later versions became larger and capable of working in deeper water. More importantly the acceptance and proliferation of the barges throughout the 1930s and 1940s led to the emergence of a marine supply and construction infrastructure in coastal Louisiana, that, once the technological problems involved with offshore drilling were solved, virtually exploded offshore.

THE LEGACY OF EXPLORATION AND DEVELOPMENT IN INLAND WATERS

The deltaic plain was not empty in the 1920s and 1930s as initial exploration and development of oil was occurring. Soon after settlers of European descent moved into the area in the mid-1700s the extraction of renewable resources from coastal wetlands became a major part of the regional economic picture. By the mid-1800s over 150 wetland-oriented communities were in existence in the Louisiana alluvial wetlands (Davis 1990). Connected only by water transportation routes, this large, isolated, but permanent population harvested shrimp, oysters, fish, furs, and a variety of other minor marsh commodities. The inhabitants of the marsh consumed marsh products and marketed them to settlements along Bayou Teche and in New Orleans via those same water routes (Comeaux 1972; Davis 1990).

Concomitant with, and aiding, the movement of oil activities into the marsh in the 1920s and 1930s was the spread of the highway network. The increasingly attractive development along the long thin ribbons of natural levees, parallel to the waterways crossing the marsh, attracted migrants from the more isolated locations in the deltaic plain and concentrated populations in long "string town" (Kniffen 1968) settlements. This abandonment of the more isolated locations in the marsh and the concentration of the population in more urban areas assisted the transition from subsistence activities in the marsh toward a more commercial harvest perspective. The growing population along the settlement strips also provided a local labor supply for expanding petroleum activities, which in turn reinforced their movement into the marsh.

Movement into marsh environs was initially accomplished by the construction of "board roads." The roads were set on a base of twelve-inch-wide by twenty-foot-long planks laid perpendicular to the direction of the road. Across the center of these planks, additional boards were laid (sometimes elevated on cross ties in the softer marsh) parallel to the direction of travel to form the road surface (see Nicholson 1942 for a detailed description). Travel across the shallow open water of the deltaic plain with heavy equipment was equally problematic, but the introduction of the submersible drilling barge was quickly followed by the means to provide access not only to the open-water bodies but also to marsh locations.

In 1938 a barge-mounted dragline prepared the first marsh location for the new class of submersible barges (McGee and Hoot 1963). Since it was usually much easier to dig canals across the marsh or through shallow estuaries, float equipment on barges to the site, and supply the operation via vessels through the newly cut canals than to prepare the marsh itself to support the weight of drilling locations and access roads, this new technology moved ahead with little regulation.

In the four decades between 1937 and 1977 approximately 6,300 exploratory wells and over 21,000 development wells were drilled in the eight Louisiana coastal parishes. As Davis and Place (1983) note, most of these were in wetlands or inland water bodies. Thus, it was during the 1930s, following the introduction of the drilling barge and barge-mounted dragline, that the extensive networks of canals associated with oil development along the northern Gulf Coast (primarily in Louisiana) was initiated. The eventual extent of the destructiveness of this practice has yet to be completely assessed (Turner and Cahoon 1988; Gagliano 1973). In lieu of virtually any permitting practices, private landholders and the state of Louisiana allowed almost unlimited access via barge to drill sites in the coastal zone, and a network of pipeline corridors through the same wetlands only exacerbated the problem. Barrett (1970) measured more than 7,300 km. of canals south of the Gulf Intracoastal Waterway in Louisiana alone, resulting in a direct land loss of over 190 square miles due exclusively to canal surface (Gagliano 1973). There is little doubt that this channelization process accompanied by erosion, alteration of the hydrologic regimes, and salt water intrusion has contributed greatly to the current annual loss of between thirty-five and fifty square miles of wetlands in the state.

As severe and irrevocable as this damage is, it can only be understood in light of the environment (both physical and social) in which it occurred. Movement into the Louisiana marsh in the 1920s and 1930s happened at a time when not only was the idea of environmental protection nonexistent (Freudenburg and Gramling 1993, 1994a), but the concept of the marsh as a valuable resource was literally unavailable. "Resources" are only those things that are valued by human cultures at a particular place and time (Freudenburg, Frickel, and Gramling 1995). Only a scant seven decades before this period, petroleum itself was not a resource because no one considered it to have social or economic worth. The same was true of the marsh in the 1930s, a hostile environment, which was seen with a "conquest of the frontier" mentality. This was an exuberant age with almost unlimited faith in technology (Catton and Dunlap 1980), and, as such, the exploration and development of oil and gas occurred in state waters as an environmentally ignorant, not malevolent, activity.

In addition, the coast of Louisiana is simply not accessible in the same way that most other coastlines in the United States are. The same band of marsh that was viewed simply as a problem by the early oil developers effectively precluded "coastal" residents from reaching the coast. The entire deltaic plain from the coast of Mississippi to almost the Texas border was inaccessible by land (and to a large extent still is). Thus, there was no conception by local residents of a "coast" and certainly not of a "beach," and this led to very different perceptions of the "value" of these "resources" by Louisiana coastal residents in the 1930s and those held in other coastal regions in the United States in the 1990s (Freudenburg and Gramling 1993, 1994a; Gramling and Freudenburg 1996 for detailed discussions of these issues).

Finally, from the time of the earliest nonnative settlers in the region, economic exploitation of the swamps and marshes had been the mainstay of local residents' livelihood (Comeaux 1972). While many of these resources (fish, crayfish, oysters, shrimp) were harvested in a sustainable fashion, others (primarily cypress lumber) were not. Although theoretically timber is a "renewable" resource in that it grows back as with the case of old growth timber, this distinction becomes rather academic, given the centuries necessary to replace it. As soon as the technology evolved to the point where massive harvest of the cypress in coastal Louisiana was possible, massive harvest is precisely what happened. By the time oil was emerging as an exploitable resource in the coastal marshes, cypress was in decline, and by the late 1930s virtually all of the old growth cypress was gone (Comeaux 1972; Norgress 1947). Oil provided an alternative, and the shift from one exploitative use of the region to another seemed natural and unproblematic[3] (Freudenburg and Gramling 1993, 1994a). Whatever the "reasons" for the initiation, canal construction has radically altered the hydrology of the northern Gulf Coast (Davis and Place 1983). Erosion and salt water intrusion have continued to the present (although attempts at control have been set in place in the last several decades) and will continue for the foreseeable future.

By the late 1940s, the technology of the submersible drilling barge had expanded in several ways. First, the barges had gotten larger and more powerful, allowing them to tap the deeper oil-bearing formations along the Gulf Coast. Drilling as deep as 18,000 feet was possible by this new class of marine rigs (Sterrett 1948a). The larger size was made possible by a second innovation:

breaking the operation into two barges with half of the drill slot built into the starboard side of one barge and the port side of the other. On site the two barges were tied together to produce a single drilling unit. In this way each of the two barges could be made the size of the older submersible rigs, effectively doubling the available space but retaining the smaller size of the older rigs for easier transportation and handling. This also meant that canals dug through the marsh did not have to be any larger than with the previous technology (*Oil Weekly* 1946b). A third innovation, which led to more flexibility and power, was the replacement of steam engines by diesel-electric rigs. The diesel engines were used to turn generators, which supplied electricity for the electric motors that powered the various elements of the drilling operation (Humble Oil, Production Department 1946).

By the early 1950s these huge inland submersibles were over 200 feet in length and 55 feet in width and were capable of drilling to 20,000 foot depths (Albright and McLaughlin 1952; McLaughlin 1953). A final introduction in 1952 of the tungsten carbide drilling bit added to the efficiency of these monstrous machines, and by 1955 inland submersibles were setting world drilling depth records of 22,500 feet (*Drilling* 1955a).

RESPONSE TO THE NEW TECHNOLOGY

The movement into increasingly isolated locations in the coastal marshes and estuaries led to the institutionalization of two additional innovations necessary for the movement offshore, one technical and one social. The isolated location of wells along the Gulf Coast with no access to roads and the necessity to move oil from wells to refineries led to the emergence of extensive pipeline networks and to the adaptation of the pipeline to the marine environment. The first real marine application came with a 25-mile stretch of pipeline crossing Lake Pontchartrain, north of New Orleans, in 1941. The solution of the logistical problems associated with positioning and assembling a continuous pipeline in a marine environment led to the early technology associated with "lay-barges" that work the Gulf today (Sterrett 1941) and the eventual proliferation of the massive pipeline network in the northern Gulf. A decade later United Gas Company was spanning over 25 miles of the open Gulf of Mexico to tie two offshore platforms into the pipeline network (Taylor 1951), and the following year the Texas Pipeline Company had completed a 22-inch in diameter 216-mile long pipeline from near the Lake Pelto field, where the Giliasso (the first submersible rig) spudded its first well, to refineries in Port Arthur, Texas (Trow 1952). Much of this line was through the coastal and marine environment.

During this same period, the necessity for construction and drilling crews to work in isolated locations, where it was impractical to commute daily, led increasingly to the use of living quarters located near the work location and the division of work into concentrated working periods (e.g., a week at work and a week off). By 1941, Nicholson (1941b: 30) was describing the luxury of the newest and largest of the remote living quarters, created from a converted river steamer and supplied by Gulf Oil near the mouth of the Mississippi, as providing "excellent food and sanitation" as well as "[f]ountains supplied with running ice water" and rooms with "reversible window fans." By the mid-1940s, the larger of these living quarters had come to be described as "modern floating hotel[s]" (*Oil Weekly* 1946c: 44).

The investment structure associated with work in the marine environment also affected work scheduling. The equipment associated with work in these environments was expensive, the ventures were capital intensive and needed to return investment as quickly as possible. These factors led to 24-hour-a-day operations and the concentrated work scheduling now common offshore (Gramling 1989).

MOVEMENT OFFSHORE

California

As noted earlier, the first drilling for oil over water occurred from piers extending into the Pacific Ocean at Summerland, California, a few miles east of Santa Barbara. Oil had been found down to the shoreline, and natural seeps in the Santa Barbara channel had resulted in pitch washing up on the beach for as long as there had been an historical record of the area. By the 1920s, the state of California was leasing offshore tracts in the channel. When applications for offshore tracts got out of hand in the late 1920s, and it became apparent that the state lacked the legal foundation to control offshore activities, leasing offshore was curtailed in 1929 by the repeal of the California State Mineral Leasing Act. In 1938 the State Lands Act was passed, giving the state the necessary legal control over offshore leasing, and limited development offshore resumed (Steinhart and Steinhart 1972). Major offshore efforts, however, did not really begin again until after World War II.

The Gulf of Mexico

In 1938, a joint effort by Superior and Gulf brought in seven producing wells in the Creole field from a platform a mile offshore from Cameron Parish, Louisiana, and in nine feet of water (Logan and Smith 1948). This was the first attempt in the open Gulf, and it used a system very similar to that practiced in Lake Maracaibo, in that the drilling was accomplished from a deck supported by pilings, and the support for drilling (crews, etc.) was primarily from the nearby shore. By the mid-1940s a number of proposals to extend the concept of the drilling barge to the open Gulf began to appear in the trade journals (Shrewsbury 1945; Tucker 1946a, 1946b). The published proposals envisioned self-contained, reusable drilling devices, illustrative of modern submersible drilling barges (Tucker 1946a), drill ships (Tucker 1946b), or even guyed production platforms (Shrewsbury 1945; Armstrong 1947), the last not actually being used until the late 1980s.

   In spite of these visionary proposals, the first attempt on what was later to be defined as the Outer Continental Shelf and the undertaking that first moved the entire drilling support system offshore relied on more traditional technology. In 1945 the state of Louisiana held an offshore sale on the Outer Continental Shelf, and the only bidder, Magnolia Petroleum Company (later to become part of Mobil), leased 149,000 acres offshore (*World Oil* 1956a). The following year, using a mixture of wood and steel pilings, Magnolia constructed a platform in the open Gulf on their newly acquired tract. The site was south of Morgan City, five miles from the nearest land. The platform was designed to withstand 150-mile-per-hour hurricane winds and a deck load of 2.25 million pounds. Drilling crews were rotated back and forth from one of the "floating hotels" anchored behind Eugene Island, ten miles away. This first endeavor foretold the future of changes in coastal occupations to come.

> For . . . [initial survey] as well as the construction and drilling operations that followed, Morgan City's shrimp fleet proved most helpful, a number of these boats manned by seamen schooled in the Gulf Coast waters, being utilized for survey purposes as well as for transportation of personnel, supplies and equipment . . . (*Oil Weekly* 1946d: 31).

Although the attempt was an economic failure (a dry hole), technologically it was a success in that it demonstrated that exploration in the open Gulf was feasible. In a development curve reminiscent of the race to the moon, the subsequent two decades that followed Magnolia's first Outer Continental Shelf attempt saw the emergence of a full-blown offshore technology capable of working in 500 feet of water and over sixty miles offshore (*Offshore* 1966a). But, before this could happen, the question of who owned offshore lands had to be settled, a question that raised a heated national debate.

THE TIDELANDS CONTROVERSY

As the technology evolved, allowing penetration into the coastal estuaries and bays of the Gulf of Mexico and the near-shore waters of the Pacific, the states of California, Louisiana, Texas, and Florida quickly realized the potential for revenue. By 1929 the state of California was issuing leases in near-shore waters, which were exploited by piers or artificial island, and in 1936 the Louisiana legislature created the State Mineral Board and directed the board to competitively lease the waters adjacent to the state. While exploration and development had not yet moved into the open Gulf, the technology existed for this step, and intensified activities in coastal Louisiana, particularly behind the barrier islands in Timbalier Bay, indicated that movement into the open Gulf was not far away. Expectations in general were high that oil would be found under the open seas.

The issue that arose, accordingly, was not one of technology, but one of ownership of subsurface lands.[4] While the states of Louisiana, Texas, Florida, and California were assuming that they owned the sea bottoms, there were those in the federal government that did not agree. Perhaps chief among these was Secretary of Interior Harold Ickes. As early as 1937, at Ickes's encouragement, a resolution passed the Senate directing the attorney general to assert federal ownership of offshore lands, but the House took no action on the matter, and World War II soon overshadowed the issue. Throughout the war, Ickes continued to push for federal ownership of offshore lands, and shortly after the end of the hostilities he approached Truman on the issue. Truman was sympathetic to federal ownership and in 1945 issued a proclamation (Executive Order 9633, *Federal Register* 12304 (1945); 59 Stat. 885) asserting federal ownership of the continental shelf.[5] In addition, Ickes persuaded Truman to initiate a suit against California in the U.S. Supreme Court. The states argued that they owned mineral deposits in the waters adjacent to their coasts as remnants of their jurisdiction over the marginal sea (the three-mile strip of sea adjacent to a nation's coast) as original colonies. The federal government argued that the marginal sea concept did not arise until after the revolution. While the suit was progressing, the issue came to the attention of the American public (as noted in chapter 1) with Truman's nomination of Edwin Pauley as under-secretary of the navy and Ickes's testimony that Pauley had offered to contribute $300,000 to the Democratic Party if he (Ickes) would get the suit stopped. In spite of the Truman proclamation and the impending suit, in the absence of clear legal precedent, the state of Louisiana (and Texas, which soon followed) continued to lease offshore lands. By 1950 Louisiana and Texas had leased almost five million acres offshore (Mead et al. 1985).

In 1947 the Supreme Court ruled against California. Ultimately, the conflicting claims to the Outer Continental Shelf were settled by this decision and a series of additional decisions between 1947 and 1950 by the U.S. Supreme Court, in what are known as the Tidelands Cases. This series of decisions established the legal rights of the federal government over all of the U.S. offshore lands (see Cicin-Sain and Knecht 1987 for a detailed discussion of this conflict). Throughout this period Congress struggled with the issue, and a number of bills were introduced to address the conflict.[6] Early in 1952 Congress did manage to pass a bill allowing state claims, which Truman vetoed.

In the wake of the Supreme Court decisions, Congressional action, and presidential veto, the Tidelands controversy became a major issue in the 1952 Presidential election, with Eisenhower supporting the "state's rights" position and Stevenson coming down clearly on the side of federal ownership. Within the oil industry, the sentiment ran heavily in favor of the states, and the specter of big government interfering in state and individual rights was a popular one.

> Under this convenient doctrine of "paramount rights"[7] of the U.S. government, it was warned, a socialist-minded administration could easily nationalize all property and socialize the country (*World Oil* 1951: 73; see also *World Oil* 1952)

The major oil companies believed it would be easier to deal with the states than with the federal government[8] and undertook a major lobbying effort in support of states' rights. Along the Gulf Coast and in California, the issue was a hot one, as Solberg notes:

> Feelings ran highest in Texas, where the Tidelands issue topped all others. When Democratic candidate Adlai Stevenson came out for Federal control in the face of crowds bearing placards "Remember the Alamo and the Tidelands Oil Steal," . . . Governor Allan Shivers announced that he could not support the party's candidate. With the hero born in Denison, Texas, leading the way, the oil of the tidelands had much to do with detaching Texas from its historic allegiance to the Democratic party. (Solberg 1976: 164)

ADDITIONAL COMPLICATIONS IN THE MIDDLE EAST

The Tidelands controversy was not the only one surrounding the issue of oil during the early 1950s. By 1950, following the close cooperation coordinated by the United States during World War II and supported by the Marshall Plan's vision to convert western Europe from coal to oil supplied from the Middle East, a tight arrangement between the major oil companies completely controlled Middle Eastern oil (Kaufman 1978; Solberg 1976). Arabian American Oil Company (Aramco) had completed a pipeline from the company's holdings in Saudi Arabia to the Mediterranean, and with prices set to what they would be if delivered from the Gulf of Mexico under the Achnacarry agreement, the companies were making huge profits on the much closer Middle Eastern oil, much of this bought with Marshall Plan funds, that is, the U.S. Treasury (Solberg 1976: 181–182).

In addition, a quiet deal had been worked out, with the cooperation of Secretary of State Dean Acheson, in writing up Saudi tax laws (none existed at the time), where the royalties paid for oil were considered taxes. Under existing laws barring double taxation and a generous ruling from the Treasury Department, this meant that the companies did not have to pay taxes on much of the enormous profit earned during this period (Solberg 1976; Senate Subcommittee on Multinational Corporations 1974 IV).

This cozy arrangement, along with the agreements several of the major multinationals had with the German government during World War II, had come under investigation by the Justice Department, and the Truman administration was trying to decide whether or not to prosecute under antitrust laws. In the midst of this, initiated by the failure to negotiate a new contract and with the rise to power of Muhammad Mossadegh, Iran nationalized British Petroleum's holdings. After failing to get compensation, British Petroleum instigated a worldwide boycott of Iranian oil. The implications of this nationalization for the remainder of the Middle East were obvious, but there was an additional consideration. With the growing cold war, keeping the Soviet Union out of the Middle East became a major U.S. foreign policy objective of the Truman administration and subsequent administrations.[9] Mossadegh, who came out of the left wing of Iranian politics, represented a perceived threat to this position. As a result the Truman administration backed the British call for compensation and the boycott. For the boycott to succeed, however, it would require the cooperative action of the major multinationals, precisely what antitrust action would contend was illegal (Kaufman 1978).

The Truman administration was snapped out of inaction caused by this dilemma with the impending release of an investigation by the Federal Trade Commission. The Federal Trade Commission had been investigating antitrust practices by various U.S. industries and in 1949 turned its attention to the oil industry. After a yearlong investigation, the detailed report documented cooperative actions by the oil companies, including the Achnacarry and Red Line agreements. Although a number of Truman administration officials urged keeping the report classified, arguing national security considerations, news of the report leaked out, forcing its release. With few options remaining, Truman allowed a grand jury investigation to go forth, sparking charges within the oil industry that the action was a "typical pre-election stunt" (Zondag 1952). Shortly before leaving office, Truman, caught in the need for cooperative action by the oil companies on the Iranian boycott, considered changing the criminal investigation into a civil suit (Kaufman 1978).

After taking office the Eisenhower administration was caught in the same bind as the Truman administration, needing cooperative action by the same oil companies that were accused of cooperative action, and moved to kill the Justice Department criminal investigation against the oil companies. The remaining civil suit died years later, but in 1953 the CIA engineered a coup to oust Mossadegh and return the Shah of Iran back to power (Wise and Ross 1954).

FEDERAL LANDS

Several months later with the threat of a veto removed, Congress passed the Submerged Lands Act (43 U.S.C. 1301-1315),[10] which assigned to the states the title to the offshore lands that were within three miles of the shoreline, and Eisenhower quickly signed it into law to the relief of the industry (Logan 1953; *World Oil* 1953a). Two current exceptions to this act involve Texas and the west coast of Florida, where the Supreme Court later ruled that the states had held title to three marine leagues (approximately 10.4 miles) as sovereign nations before they were admitted to the

Union. Although there were challenges to the Submerged Lands Act by states that did not stand to benefit (*World Oil* 1953b, 1954a), conflicts between coastal states and the federal government[11] (Logan 1953) and even between departments within the government (*World Oil* 1954b) over just where state waters ended, the Submerged Lands Act eventually defined the limits of state ownership.

Ironically, while much attention was devoted to the states' rights issue, very little controversy centered on the submerged lands beyond the three-mile limit. Because of the Supreme Court decision and in the absence of Congressional action deeding these lands or a portion of the revenue from them to the states, they soon passed totally, securely, and with little conflict into federal ownership. Why this happened is not clear. By 1946 Magnolia Petroleum had demonstrated that drilling was possible beyond three miles offshore (*Oil Weekly* 1946d), and by 1948 there were twenty-four operations underway over three miles offshore in the Gulf of Mexico (Logan and Smith 1948). The following year, technology had evolved to the point where a mobile offshore drilling rig was available and at work. Shortly thereafter offshore drilling was known to be feasible in over sixty feet of water (Logan 1953), and industry reports were touting the resource potential of the Outer Continental Shelf (*World Oil* 1953c, 1954c). The states of Louisiana and Texas had already sold over 300 leases in the Gulf over three miles offshore (Minerals Management Service 1993), most off the Louisiana deltaic plain, and there was little doubt that the offshore lands would produce oil.

The answer may lie in the fact that the action for federal ownership was initiated by litigation, which, as Gramling and Freudenburg (1992a) have noted, tends to lead to intractable positions on both sides. Further, the legal battle came to be over the "Territorial Seas," which as a matter of international law had come to be defined as out to three miles (Cicin-Sain and Knecht 1987). Thus, it may well have been not ignorance of the potential of these lands, and certainly not by those most directly involved in the industry and in the states of Louisiana and Texas, but rather lack of conceptual clubs with which to enter the legal-political fray, coupled with a rush by the Eisenhower administration to make good on campaign promises that led to the limiting of state lands to three miles. Whatever the cause, if the intent was to hand over the bulk of offshore resources to the states, the action was a miserable failure.

The second piece of offshore legislation, which ultimately was of much more importance, was the Outer Continental Shelf Lands Act (OCSLA) of 1953 (43 U.S.C. 1331–1356). This act authorized the secretary of interior to lease, through competitive bidding, the offshore lands ("out") beyond the three-mile limit, the "Outer Continental Shelf" for the development of the oil, gas, salt, and sulphur resources found there and to subsequently administer these leases. The act served as the major legislation guiding policy for federal offshore leasing starting in 1953 until it was amended in 1978.

Within the Department of Interior, the Bureau of Land Management (BLM) and the U.S. Geological Survey (USGS) were given the shared responsibility for this task. Initially, after conducting a resource evaluation of a broad offshore area, an invitation to nominate tracts within that area was published in the *Federal Register*. Selection of a number of offshore tracts (each limited to a maximum of 5,760 acres, a three-mile-square "block") was based on the ensuing indication of interest by industry and on the resource assessment conducted by USGS. Eventually a sale date and a final selection of tracts was also published. Although the OCSLA allowed some leeway, the Department of Interior prior to 1978 generally used a cash bonus system with a fixed royalty rate of 16 3/4 percent. This meant that bidders bid a cash amount to lease a tract and agreed to pay royalties of one sixth of any subsequent production (see Mead et al. 1985, for more detail).

The first Outer Continental Shelf lease sale was held in the Gulf off Louisiana on October 13, 1954. Ninety tracts were leased (Gould et al. 1991), and an aggressive Outer Continental Shelf development program was underway. Later that year, a lease sale was held off Texas, and by 1959 a lease sale had been held off Florida.[12] Although Louisiana and Texas went to court to block the implementation of the OCSLA and succeeded in delaying offshore sales in 1956, 1957, and 1958, ultimately the states lost, although Louisiana continued the fight until the mid-1960s (*Offshore* 1966a), resulting in sporadic lease sales during this period. With these early exceptions, lease sales in the Gulf have been held at least annually and as frequently as five times a year (in 1974) (Gould et al. 1991). Thus, with the settlement of the ownership of offshore lands and the initiation of the federal leasing program, offshore development moved ahead rapidly, at least in the Gulf. The rapid evolution of offshore technology and the systems to support it are examined in chapter 3.

Chapter 3

Moving Offshore

EMERGENCE OF THE OFFSHORE SYSTEM: TECHNOLOGY

Between 1946 and the settlement of the ownership issue in 1953, offshore development proceeded, albeit at times in a legal vacuum. Following on Magnolia Petroleum Company's lead and using the same type of technology, Kerr-McGee brought in the first producing well on the Outer Continental Shelf in 1947, twelve miles off the Louisiana coast in eighteen feet of water. This sparked additional interest offshore, and by the middle of 1948, in addition to offshore production by Kerr-McGee and the Creole field initiated in 1938 (see chapter 2), thirteen locations were in various stages of development off the coasts of Louisiana and Texas, two additional dry holes had been abandoned, and plans were underway for drilling in fourteen other locations (Logan and Smith 1948). Magnolia Petroleum Company and Kerr-McGee had emerged as the clear leader for continental shelf operations (Logan and Smith 1948; Seale 1948). One of the more promising prospects underway in 1948, by Humble (later Exxon) off Grand Isle, Louisiana, resulted not only in significant discoveries but also in the establishment of a support base on Grand Isle (Sterrett 1948b), which has remained in continuous operation to the present. With a success rate offshore of 70 percent (*World Oil* 1956a), almost 200 structures were in place on federal waters by the time the first official federal offshore lease sale was held in 1954 (Carmichael 1975).

All of these structures were in relatively shallow water, and many of them were simply platforms for exploratory drilling. A short two years later *World Oil* would proclaim the emerging offshore development "successful," and paint a rosy picture for future offshore development (Cram 1956: 78).

The basic technological problem faced throughout the early development offshore was essentially the same as that faced in the marsh, prior to Texas Company's introduction of the mobile drilling barge in 1933. Because drilling took place from a fixed platform, the cost of each offshore attempt was between $750,000 and $2 million, a considerable sum in the 1940s, all sunk cost and lost in the event of a dry hole. Most offshore efforts had quickly moved to smaller (less expensive) platforms to house the drilling machinery and floating tenders moored to the platform, which contained crew quarters and storage. The tenders were at least mobile and reusable. Mobile drilling barges used in protected waters, which had reached considerable size and complexity by this time (Sterrett 1948a), were not appropriate for use offshore because of water depths and because wave action could move them laterally during storms. Platforms, which had by this time withstood hurricane-force winds (Farley and Leonard 1950), derived their strength not only from the fact that their pilings were driven deeply into the bottom, a firmer bottom than that found in the marsh and estuaries, but also because waves could pass through the pilings.

The breakthrough, equivalent to Texas's Galliaso, came in 1949 when John Hayward designed an offshore drilling barge. The barge was a 160-by-54 foot compartmentalized barge with a drilling slot similar to those used in protected waters. But the drilling platform, machinery, and crew quarters were mounted on a platform twenty feet above the deck of the barge, supported by a latticework of braced columns or posts (leading later vessels of this type to become known as "posted barges"). In practice, the barge was brought to the location and the compartments in the barge were flooded in a controlled fashion to maintain stability. Resting on the bottom, the barge provided a stable base for the drilling platform. Because wave action could pass through the columns, as with pilings on an offshore platform, the barge was not moved off of its location during storms. As this top-heavy structure was inherently unstable during the flooding procedure, two large pontoons, running the length of the barge, were attached to columns on either side of the barge. These pontoons remained floating (providing stability) until the barge was firmly on the bottom, and then they too were flooded, adding additional stability on the bottom. Raising the barge involved reversing the procedure, floating the pontoons first for stability before the main hull was raised. When the Bretton Rig 20 (as the first barge was named) moved off its first location in the Gulf and on to the second location (there was considerable skepticism that the barge would ever be successfully refloated), the "revolutionary" nature of the barge was noted (Wolff 1949). All of the previous technological innovations (mobility, transparency to wave action, offshore scheduling of work, and the inclusion of living quarters) came together for the first time in the Bretton Rig 20 to produce the first self-contained, reusable, offshore drilling machine.

Throughout the early 1950s, in spite of the success of the Bretton Rig 20, the offshore drilling picture was dominated by fixed platforms with floating tenders (*World Oil* 1954d). A number of innovative procedures were used to try to increase the speed and efficiency of offshore drilling, including installing two drilling rigs on one platform, doubling the drilling speed but not doubling the costs, as the tender and much of the support equipment could be shared (Kastrop 1950). Submersible rigs became the dominant type of mobile offshore exploratory drilling rig during the decade following their introduction, although the three alternatives to submersibles (jackup rigs, drill ships and barges, and semisubmersible rigs) all appeared in the 1950s and early 1960s.

GROWTH OF THE OFFSHORE SYSTEM: TECHNOLOGY

Exploration

By the time of the first federal sale in 1954, there were four submersible rigs, similar to and including Bretton Rig 20, operating in the Gulf. These had been joined by the first jackup rig and one floating drilling barge, which was only used to take cores (geological samples) for survey work (Howe 1966c). All exploratory drilling that was not done by the five exploratory rigs was accomplished by the traditional technique, initiated by Magnolia Petroleum, of constructing a platform in place. The basic problem with the submersible rigs of the Bretton 20 design was the limitation as to how high above the barge the drilling platform could be mounted and still retain stability while afloat. As a result, there was a limit to the water depth in which this design could work. Kerr-McGee did commission and put into operation a larger barge late in 1954 capable of operating in up to forty feet of water (Gibbon 1954), but that appears to be about the practical limit of the design. There were also developments in drilling from fixed platforms, the most notable being the construction of modular units that could be loaded onto completed platforms and bolted together to begin drilling. By the mid-1950s the entire offshore drilling machinery structure was available in four units ranging from 30 to 184 tons. Each of these could be lifted by crane onto a completed platform and bolted together. The time needed to begin drilling was measured in hours. Still, the limitations of the fixed platform remained the initial cost, and with water depth being the restriction of the submersible barge (fixed platforms had been set in up to 100 feet of water), four new designs emerged during the 1950s and early 1960s that addressed these limitations.

The first of these was the jackup rig. This design, which was tested in 1953 (*World Oil* 1953d), first appeared in 1954 (Howe 1966b) with prototypes theoretically capable of use in waters up to 100 feet by 1955 (Davidson 1955). The arrangement consisted of a barge on which the drilling equipment, machinery, and crew quarters were mounted. The barge was towed to the drilling site, and legs along the periphery of the barge were extended to the bottom, lifting the barge clear of the water, above wave action. Drilling occurred either through an opening in the barge or over one of the sides. The first jackup tested was a standard 300-foot barge with six 100-foot-long legs on each side (*World Oil* 1953d), but by 1955 the design had evolved to the tripod arrangement in use today. With this design, first evidenced by Zapata Off-shore's Rig No. 1, three legs, each consisting of several vertical pipes braced with a latticework of smaller tubular braces, were positioned to lift a much lighter triangular platform (Hanna 1955). Although the three-legged structure was assessed as "unusual in design" while it was under construction (*Drilling* 1955b: 68), the design became the industry standard following its successful deployment off Aransas Pass in 1956 and the survival of 100 mile-per-hour winds from Hurricane Audry by a similarly designed rig off Cameron, Louisiana, in 1957 (*Drilling* 1957).

In fact, an interesting insight on the rapid rate of the evolution of offshore technology can be made by a comparison of industry comments on the design. The first tripod Zapata jackup, which received the "unusual" comment in 1955, was built by R. G. LeTourneau in 1955. Three years later, in 1958, when the same firm unveiled a much longer-legged version it was described as "similar in appearance to the *conventional* LeTourneau design" (*Drilling* 1958a: 84 emphasis added). Following the longer-legged tripod version, jackups have evolved into massive structures that today can generally work in up to 300 feet of water.

The second innovation in offshore drilling was the drillship. Introduced in 1953, the first drillships were converted offshore barges or small ships with the derrick mounted over the bow (barges) or over the side (ships). A number of vessels were converted in 1955 and 1956 for core drilling to obtain geological information in the Pacific off California and by 1957 were drilling in over 1,000 feet of water (*World Oil* 1957a). By the early 1960s, drillships were being built from the ground up with center-line derricks for drilling through the center of the vessel and with dynamic positioning equipment replacing anchors. The third innovation occurred in 1956 when Kerr-McGee introduced the "bottle" type submersible drilling rig. The drilling platform was mounted on the top of "bottles" (large cylindrical bottle shaped tanks, similar in shape to a wine bottle, which were connected by a latticework of pipe and positioned around the periphery of the rig). When empty, the bottles floated the platform high over the water. Towed into position, the bottles were flooded, and their bottoms sat on the sea floor to provide a stable base for drilling. Because the bottles were mounted around the periphery of the drilling platform and even when empty were half-submerged from their own weight and the weight of the drilling platform, the design was very stable. Since the platform sat on the narrow necks of the bottles, the narrow necks offered little resistance to wave motion when sitting on the bottom. This design ultimately resulted in rigs capable of drilling (with 25-foot deck clearance) in 175 feet of water (Howe 1966a).

Finally, in 1962 semisubmersible drilling rigs were first used. Early semisubmersibles were converted bottle-type submersibles that were partially submerged for stability in deep water and anchored in place. Later designs involved multiple, parallel, submerged cylindrical hulls (like submarines) connected together and to a drilling deck with a latticework of shafts. Because the majority of the weight was below the water line, the rigs were very stable and because the only surfaces subjected to wave action were the columns connecting the hulls to the deck, this design was able to operate in rough water. These later designs are also self-propelling and like modern drillships are capable of transporting themselves to virtually any region of the world.

Although there were a number of proposals for other types of offshore mobile rigs, such as totally submerged drilling rigs that operated below the surface (Martin 1956), early forerunners of the gravity-based platforms used later in the North Sea (Gregory 1955), and even innovative uses of existing technology, such as transporting offshore and using a submersible drilling rig designed for inland waters from a converted dry dock (*Drilling* 1956a), the four standard designs came to dominate the U.S. offshore picture during the 1950s and 1960s. Jackups continue to be the most common offshore rigs with semisubmersibles emerging as the choice for rough water and drillships the choice for greater water depths.

In addition to the technological changes in the design of the rigs, there were corresponding changes in the drilling operations themselves and in support facilities. Marine radar and radio emerged as commercial products following World War II and were promptly adapted by the offshore support and communications sector. The concept of moving the drilling rig to different locations on the offshore platform in order to drill multiple wells without moving the platform emerged almost concurrently with the federal leasing program (*Drilling* 1956a). Safety regulations for offshore structures were first promulgated by the U.S. Coast Guard in 1956 (*Drilling* 1956b).

By 1966, a little over a decade after the federal offshore leasing program commenced, there were an estimated 150 mobile drilling rigs in operation, most of them in the Gulf (Howe 1966a, 1966b, 1966c). Most of the continental shelf in the Gulf, at least out to 500-foot depths (*Offshore* 1966a), was open for exploration. Submersible rigs ceased to be built by the late 1950s. Although jackups, drillships, and semisubmersibles have continued to evolve, the basic designs were in place by the mid-1960s, and some of the major offshore drilling companies had begun to dominate the offshore picture, as well as specialize in the type of rigs they were known for. Global Marine was the dominant drill ship company in the world, with Ocean Drilling and Exploration Company (ODECO) capturing the recently emerging semisubmersible market, and The Offshore Company, Penrod Drilling, and Zapata Offshore dominating the deep-water jackup market.

While the rate of technological development in offshore rigs between 1955 and 1965 was astounding, it was not without mishaps. During this same period there were twenty-three "major mishaps" (Howe 1966a) involving the capsizing and/or sinking of offshore rigs. The early jackup rigs were the most vulnerable. Six of them were lost in 1965 alone.

Production

All of the types of mobile drilling rigs are used during the exploratory phase of development or to later drill additional wells on existing platforms. Once oil or gas is found by exploratory drilling and the extent of the field has been delineated by more drilling, a production platform must be put in place. The production platform provides a stable base to drill production wells from and to mount the necessary equipment to control production from the wells. Early production platforms were the same platforms constructed for exploratory drilling and, of course, were built in place (i.e., piles were driven into the sea bottom and a deck was constructed on top of them). These platforms also provided mooring for barges to take the crude oil to onshore processing facilities. Both the on-site construction of platforms and the transportation of oil by barge began to change by the mid-1950s.

Production offshore in the Gulf underwent a transformation similar to that for exploration in the decade following the first federal lease sale. The most fundamental change came because it was not practical to build production platforms in place in deeper water. Because the pilings could not be effectively braced below the water line after they were driven, sufficient stability to support the weights and vibration involved in offshore drilling could only be accomplished in relatively shallow water. This meant that for deeper water, the entire production platform had to be fabricated on land and set in place offshore. By 1955, one year after the first federal lease sale and six years after the first submersible drilling barge, a number of these "self-contained" platforms (because they did not require tenders during drilling) were in place in the Gulf (Hanna 1955). That same year, the first fabrication yard primarily for the construction of offshore platforms, the McDermott facility, opened east of Morgan City, Louisiana, on Bayou Boeuf. Eventually this facility expanded to over 1,000 acres (Davis and Place 1983). It is still in operation today.

The technology for the production and installation of these platforms developed quickly. Also in 1955, McDermott produced a marine crane capable of lifting 800 tons, touted as the "largest single mechanical lift ever created" (*Drilling* 1955c: 93), for use in the installation of offshore platforms. McDermott had installed platforms in 100 feet of water late that same year and in 200 feet of water by 1959 (Clark, 1963).

Platforms are designed for a specific location and offshore field and are ordered after exploratory drilling and the characteristics of the field are known. In the United States, the "steel jacket" production platform is the norm. Early platforms were simply rectangular structures consisting of a series of parallel upright columns, supporting a deck and connected by a latticework of smaller pipes. As production moved further offshore by 1956, the basic design of the jacket (that portion that extends from the sea floor to above the water line) had been modified to resemble a slender truncated pyramid still constructed of a latticework of upright columns and braces but with the base being wider than the top of the structure to provide increased stability (*World Oil* 1956b). The size and configuration of the jacket is based on the depth of the water, bottom support condition, the number of wells, and the processing equipment required. Whatever the size, the jacket and the deck (the portion of the platform above the water line, with all the production equipment, living quarters, etc.) are constructed separately. After the jacket is barged offshore[1] and secured in place by pilings driven through the vertical columns, the deck is brought out and installed. While the description implies a simple operation, steel-jacketed platforms have been installed in up to 1,350 feet of water. This ("Bullwinkle") platform is 1,615 feet tall[2] and 400 feet across at the base.

Transportation and Supply System

Through the 1950s most offshore operations in the Gulf brought oil ashore via barges, and for a while there was considerable interest in offshore submersible storage tanks (Lacy 1957; Lacy and Estes 1960). As oil and gas production moved further and further offshore, however, so did the pipeline network. In 1951, the first large-diameter pipeline for offshore production was laid in the Gulf (Clark 1963), and by 1958 the development offshore had reached the point that the construction of underwater gathering lines was beginning to be economically feasible (Edmondson 1958). Pipelines are laid from a specially designed "lay barge." The lay barge has an anchoring system, which allows it to be pulled across the area that the pipeline is to traverse. As it moves over the area, sections of pipe are welded to the end of the pipeline, coated to prevent rust and marine growth, and are passed along a ramp that is built into the barge, and over the stern of the vessel. Once in the water, the pipe rides along a supporting ramp, or "stinger," to the bottom where it is buried under the sea floor. As the pipeline networks moved into deeper water, the lay barges became larger and more sophisticated. Like offshore drilling, the lay barge constitutes a system that must be constantly supplied from shore with sections of pipe, coating material, a rotating labor supply, fuel, food, and water.

The earliest offshore attempt, on what is now federal waters, by Magnolia Petroleum in 1946 used local expertise and vessels (primarily shrimp boats) for transportation of personnel and supplies. Following World War II, Navy LSTs became available as government surplus, and these were quickly purchased by a number of offshore operators and pressed into service as offshore tenders

(Hanna 1955). Concurrent with the spread of offshore activities, methods of supply evolved quickly, and, as with mobile drilling rigs, the prototypes for offshore vessels were soon apparent. Some of the necessary vessel types for offshore operation, such as marine tow boats (for moving exploratory rigs and steel jackets on barges) were already available, but others quickly materialized. The two vessels most closely associated with the offshore industry are the "crew boat" and the "supply boat," both of which appear to have evolved from their military equivalents, "PT" boats and "LSTs." By 1956 crew boats and supply boats built specifically for offshore oil and gas activities were becoming the norm (Craig 1956; *World Oil* 1957b).

Crew boats are fast-planing hulls; originally wooden and 50 to 60 feet long, they are today usually constructed from aluminum and are up to 150 feet long. Although, as the name implies, they were first commonly used to transport crews offshore, as helicopters gradually took over that role they are, today, more commonly used to transport specialized equipment and smaller quantities of supplies that do not merit a supply boat. Supply boats are larger displacement[3] hulls with raised bows for crew quarters and low flat decks for carrying heavy loads (such as drill casing or cement) extending over the back two thirds of the vessel. They frequently have built-in tanks below deck for water or drilling mud. Typically in the Gulf of Mexico, supply vessels are 180 to 220 feet in length, although in some areas of the world, such as the North Sea, they are frequently over 250 feet in length (Alderdice 1969).

By 1955, a year after the first federal lease sale on the Outer Continental Shelf, the helicopter had made its appearance (Monroe 1955), and it quickly became an integral part of the offshore transportation picture (Hanna 1957). With the rapid expansion offshore in the 1960s and early 1970s and the increasing distance from land involved in the movement into deeper waters, helicopters became the standard means of transporting crews and specialized personnel offshore. The expansion of all of these forms of transportation created a burgeoning transportation sector in the northern Gulf of Mexico with the construction and leasing of vessels and leasing of helicopters constituting a major economic activity in their own right. By the late 1970s, Petroleum Helicopters based in Lafayette, Louisiana, had become the world's largest helicopter company.

Far more than with most industrial sectors, transportation is an integral part of offshore oil and gas activities and that necessary mobility has a number of implications that emerged early in the Gulf of Mexico.

Mobility

In contrast to the more geographically specific types of development (a generating plant is built, a mine is opened, etc.), offshore energy exploration and development tends to be a highly mobile phenomena, and this has major implications for areas in which it develops. Not only does this mobility tend to disperse both the positive and negative social and economic impacts associated with its activities, but, as we will see, if economic or resource availability factors dictate, much of the offshore sector can relocate. This mobility is evidenced in four basic areas. First, the development itself is highly mobile. Offshore drilling rigs can be, and frequently are, moved to virtually any coastal area in the world. The exploratory drilling phase, through the development drilling and completion phases of offshore activity, is the most labor and resource intensive, leading to the majority of the employment associated with Outer Continental Shelf activity. During this time, drilling activities offshore must be supplied with labor, drilling mud, casing and drill pipe, fresh water for drilling, food and potable water for human consumption, fuel, and a variety of specialty tools and equipment. It is also during this period that platform construction will take place, and pipeline to connect the platform to the existing network will be laid. Once the wells are tied into pipeline networks, the actual production of petroleum offshore and the maintenance of platforms is much less labor intensive and provides considerably fewer jobs. Thus, the basic economic activity upon which development hinges is a highly mobile phenomenon. Additionally, the primary support sectors must also shift the concentration and delivery pattern of their products to follow development.

A second factor in the mobility of the offshore energy production industry is the transportability of many of the products that the industry and support sectors buy. While the transportation of products is characteristic of much industrial development, with the exception of the transportation industry (shipping, railroads, trucking), rarely are the construction projects associated with that development transportable. Three types of construction projects are commonly purchased and transported by the various industrial sectors associated with offshore energy production: drilling rigs of various types, production facilities (jackets, platforms), and support vessels. Since these products must be constructed on land and transported to a site offshore, they are inherently mobile. This means that they can be produced in any coastal region in the world. Thus, the construction or fabrication associated with offshore development does not necessarily have to occur in the vicinity of that development. While extensive local fabrication has been associated with offshore development in Louisiana and Texas, this occurred because of the history of that development (Freudenburg and Gramling 1994a).

In the early days of offshore exploration, Louisiana was the primary area of offshore activity in the world. As activities moved farther and farther offshore into new environs, this required modification and redesign of existing equipment. Because the need was local and for local conditions, this allowed the development of a strong fabrication and shipbuilding sector of the economy with little competition from a world market. Recently, however, areas of the world with much lower labor cost have become extremely competitive in the production of offshore drilling rigs, platforms, and vessels, and this trend will undoubtedly continue for the foreseeable future. When planning was underway for the development of the massive Hibernia field off Newfoundland, there was considerable debate, leading to intervention of the provincial government, as to whether the production platform would be a gravity-based[4] concrete platform that would be constructed locally or a floating fabricated-metal platform that would probably have been built in Asia. Pressure from the government helped to turn the tide for a gravity platform and local jobs.

Third, the products themselves, oil and gas, are also very mobile. With few exceptions, production platforms in the United States are tied directly into pipeline networks, and the products are brought ashore via pipeline. As pipelines can be run practically anywhere, secondary treatment, refining, scrubbing, and so on, can also take place far from the actual recovery activities. The exception to this, off-loading directly to tankers, actually provides an even more mobile alternative.

Taking all three of these factors into account means that the impact of offshore activities, jobs, and economic development may or may not accrue to the communities closest to that development. Thus, offshore activities in many cases act as a distribution mechanism, distributing both positive (jobs, income, and other economic activities) and negative (stress on community infrastructure, family problems) impact far beyond the geographic area in which primary development occurs. This leads to a fourth and final type of mobility, that associated with the labor force employed in offshore and related sectors, which is discussed in detail under "Social Environment" later.

DEVELOPMENT OF THE ONSHORE SUPPORT SYSTEM

Infrastructure and the Physical Environment

There are a number of factors that shaped the evolution of the infrastructure supporting offshore development in the Gulf. Because the system for offshore oil and gas production evolved in the Gulf of Mexico, there was no precedent, and the equipment and structures used for offshore exploration and development had to be locally constructed (often from the ground up) to meet local needs. Moreover, because of the rapidity with which the industry moved offshore during the 1950s and early 1960s, design and engineering were in a constant state of flux. The result of these factors was that, unlike areas of the world where offshore production came later and where because of past experience planning was at least possible, if not always successful (Manners 1982), the onshore support system in the northern Gulf emerged as a virtually unplanned entity.

An additional factor was the nature of the physical environment itself (see prologue). The deltaic plain of Louisiana from where most of the activity was supported was aptly described by Davis and Place.

East of Vermilion Bay, sediment accumulation has resulted in the sequential development of a delta system. Known as the deltaic plain, the area is the site of a series of six major deltaic lobes that extended seaward at different times during the last 7,000 years. Each lobe advanced into the shallow waters of the Gulf of Mexico . . . and was distinguished by numerous distributaries. . . . These channels continued to bifurcate aiding the distribution of the river sediments and progradation of the coast. Natural levees formed along the channels and served as favorable settlement sites.

Two distinctive plant communities have influenced development of the deltaic plain by the oil and gas industry—flotant and roseau cane. . . . Flotant communities, distinguished by abundant "plant species tolerant of frequent and sustained flooding," are "anchored in a relatively thin, matted layer of decomposing vegetable debris that is either truly floating on water or supported by highly aqueous organic ooze" (Russell 1942, pp. 78–79). Roseau cane marsh communities, on the other hand, are dominated by tall grasses and reeds and essentially is land; it can support a man's weight . . . while flotant "trembling prairie" is a quagmire underfoot. The ease of excavation of flotant marsh has facilitated development of networks of canals associated with the 144,900 ha [approximately 560 square miles] classified by the USGS as extractive land, a category almost synonymous with the oil industry in coastal Louisiana. (Davis and Place 1983, pp. 4, 8)

The nature of the terrain restricted settlement and industrialization to the many scattered natural levees of current or previous waterways, and while at times high ground was at a premium, access to the Gulf certainly was not. Anywhere one of these strips of high land met an existing waterway was a potential construction or support site, and these sites sprung up like mushrooms as offshore development moved ahead. The result was that the emerging support, transportation, and fabrication sector was scattered across southern Louisiana and later Texas. Fabrication yards ranged from the 1,000+ acre sites near Morgan City to operations the size of half an acre with a welding truck and a port-o-let. Staging areas similarly ranged across the coast, and concentrations quickly materialized at Venice, Bayou Lafourche, Morgan City/ Berwick, Intracoastal City, Cameron, Orange, and Port Authur. Wherever road or rail met waterways, local docks and staging areas appeared.

Some of these areas, like Morgan City/Berwick and the towns along Bayou Lafourche, were fishing communities before the turn of the century. The towns grew rapidly, often with little planning. Existing settlement patterns were altered, transportation networks were extended, and community services were quickly overwhelmed. Morgan City became a textbook example of rapid community expansion (Stallings et al. 1977; Manual 1980; Gramling and Brabant 1986a) with housing shortages, inadequate utility networks and social services, and an economy almost totally tied to offshore support (Gramling and Freudenburg 1990).

Other areas like Venice and Intracoastal City had little in the way of indigenous populations in the 1950s and as a result became commercial settlements with staging areas and company offices but little in the way of a permanent population. Lafayette, centrally located and able to offer more amenities, became a corporate headquarters city for the offshore industry during this period (Gramling 1983; Gramling and Brabant 1986a). Coastal Texas, which had already been transformed by the oil industry, received new impetus during this period with Houston becoming the financial center for oil-related activities (Feagin 1985) and major fabrication yards springing up in the Beaumont, Orange, and Port Arthur areas.

By the early 1960s, over 100 platforms a year were being placed offshore in ever-deeper water (see Table 3.1). This meant that not only were over 100 platforms fabricated each year and moved offshore and supplied while production drilling was proceeding, but also that each year an additional 100 platforms were added to the growing base of offshore facilities that had to be serviced and maintained.

Through the analysis of telephone directories, Davis and Place (1983) identified over 3,500 businesses in coastal Louisiana alone, directly serving the petroleum industry by the early 1970s. This was up from approximately 1,200 businesses in the mid 1950s, a growth rate of over 100 businesses a year. These firms were located on over 28,000 acres of land given over to industrial and

Table 3.1. Installation and Removal of Production Platforms on the Federal Outer Continental Shelf

YEAR	GULF OF MEXICO		PACIFIC	
	INSTALLED	REMOVED	INSTALLED	REMOVED
1953–1960	474	7	0	0
1961	111	0	0	0
1962	123	0	0	0
1963	86	2	0	0
1964	130	0	0	0
1965	126	5	0	0
1966	116	2	0	0
1967	134	0	1	0
1968	107	1	3	0
1969	105	1	1	0
1970	124	1	0	0
1971	101	4	0	0
1972	121	6	0	0
1973	98	1	0	0
1974	52	18	0	0
1975	100	32	0	0
1976	111	31	1	0
1977	116	15	1	0
1978	164	25	0	0
1979	159	43	2	0
1980	169	32	3	0
1981	171	24	2	0
1982	192	21	0	0
1983	177	36	1	0
1984	229	47	1	0
1985	214	68	3	0
1986	109	33	1	0
1987	119	24	1	0
1988	178	102	0	0
1989	188	100	2	0
1990	174	107	0	0
Total	4,578	788	23	0

Source: Gould et al., 1990; American Petroleum Institute 1993; Minerals Management Service 1993.

commercial uses scattered throughout the coastal zone of the state. If all of this marine activity and the associated support and managerial activity had been concentrated into one port, it would have been by far the busiest commercial port in the world from the late 1950s through the mid-1980s.

The Push For Development

By 1957 there were 101 offshore drilling rigs operating in the United States, 96 of these were operating off Louisiana (with 5 off Texas and 1 off California), most of them off the deltaic plain (*World Oil* 1957c), when an increasingly active federal leasing program began to kindle what would become an offshore boom. Table 3.2 shows production, leasing, and exploration and development well trends on the Gulf Outer Continental Shelf from 1954 through 1990.

Starting from the 275 tracts leased by Louisiana and Texas prior to 1954, a total of 237 tracts were offered during the first year of the Outer Continental Shelf leasing program (1954), and of these, 109 (46%) were leased. As can be seen from Table 3.2, there were a number of years in the late 1950s and early 1960s when there were no lease sales in the Gulf. However, the stipulations associated with the leases served to smooth the offshore development curve and in general drive it steadily upward. The primary term of the lease was for five years,[5] but the lessee could retain the lease beyond this period if the tract was producing or, with approval of the Department of Interior, as long as drilling operations were underway (43 U.S.C. § 1337(b)). Thus, once a lease was obtained, the lessee at least had to have exploratory drilling underway within five years or risk losing the lease. Because of limitations on available offshore rigs, the lag time involved in planning offshore operations and the fact that the major players held many leases and had to balance their attention between them, there was no one-to-one annual correspondence between leases sold and offshore activity. There was, however, a pressure to begin exploration on a lease sometime within five years in order to keep the lease, and if commercially feasible quantities of oil or gas were found, then there was financial incentive to bring the tract into production. In

Table 3.2. Gulf of Mexico Outer Continental Shelf Activities

YEAR	TRACTS OFFERED	ACRES OFFERED	TRACTS LEASED	ACRES LEASED	TOTAL BONUSES COLLECTED
1954	237	860,608	109	461,870	139,735,505
1955	210	674,095	121	402,567	108,528,726
1956	0	0	0	0	0
1957	0	0	0	0	0
1958	0	0	0	0	0
1959	118	539,813	42	171,300	89,746,992
1960	385	1,610,254	147	704,526	282,641,815
1961	0	0	0	0	0
1962	830	3,718,115	420	1,929,177	489,481,061
1963	0	0	0	0	0
1964	28	34,028	23	32,671	60,340,626
1965	0	0	0	0	0
1966	70	263,891	41	139,773	188,010,893
1967	206	971,489	158	744,456	510,079,178
1968	195	775,375	126	570,983	743,767,835
1969	65	190,153	36	108,657	110,945,535
1970	161	666,845	138	598,510	945,064,773
1971	18	55,872	11	37,222	96,304,523
1972	210	970,711	178	826,195	2,251,347,556
1973	276	1,514,940	187	1,032,570	3,082,462,611
1974	1,006	5,006,881	356	1,762,158	5,022,860,815
1975	1,143	5,989,734	265	1,369,828	670,821,011
1976	193	942,092	77	337,413	555,125,455
1977	223	1,074,536	124	605,427	1,170,093,432
1978	362	1,865,423	206	1,052,467	1,666,298,621
1979	247	1,166,118	171	812,702	3,160,826,960
1980	273	1,367,883	183	934,977	4,094,889,184
1981	421	2,159,295	258	1,308,213	3,893,097,504
1982	378	1,952,417	171	871,478	1,802,832,942
1983	13,023	71,153,488	1,040	5,393,997	4,906,889,551
1984	20,816	115,413,886	970	5,125,178	2,478,473,398
1985	15,754	87,028,709	670	3,529,325	1,542,346,514
1986	10,724	58,670,104	142	734,427	187,094,747
1987	10,926	31,846,415	640	3,447,825	497,247,006
1988	11,282	61,492,451	917	4,829,523	514,083,346
1989	11,013	60,097,672	1,049	5,580,867	645,646,870
1990	10,459	56,788,766	825	4,263,446	584,301,918

(continued on next page)

Table 3.2. continued

YEAR	$/ACRE COLLECTED	EXPLORA-TORY WELLS	DEVELOP-MENT WELLS	OIL PRO-DUCTION[a]	GAS PRO-DUCTION[b]
1954	303	3	61	3.3	56.3
1955	270	18	117	6.7	81.3
1956	NA	41	182	11.0	82.9
1957	NA	47	282	16.1	82.6
1958	NA	56	172	24.8	127.7
1959	524	86	213	35.7	207.2
1960	401	114	259	49.7	273.0
1961	NA	110	335	64.3	318.3
1962	254	148	351	97.4	451.9
1963	NA	188	357	104.6	564.4
1964	1,847	200	474	122.5	621.7
1965	NA	169	619	145.0	645.5
1966	1,345	260	596	188.7	1,007.4
1967	685	287	604	221.9	1,187.2
1968	1,303	294	651	269.0	1,524.1
1969	1,021	215	607	312.9	1,954.5
1970	1,579	201	628	360.6	2,418.6
1971	2,587	254	556	418.5	2,777.0
1972	2,725	296	545	411.9	3,038.6
1973	2,985	291	527	394.7	3,211.6
1974	2,850	336	459	360.6	3,514.7
1975	490	310	517	330.2	3,458.7
1976	1,645	278	782	316.9	3,595.9
1977	1,933	322	851	303.9	3,737.8
1978	1,583	305	808	292.2	4,385.1
1979	3,889	334	753	285.6	4,672.9
1980	4,380	349	754	277.4	4,641.4
1981	2,976	327	808	289.7	4,849.5
1982	2,069	372	792	321.2	4,679.5
1983	910	378	717	348.3	4,040.7
1984	484	559	710	370.2	4,537.8
1985	437	490	617	368.3	4,000.9
1986	255	263	396	389.2	3,948.9
1987	144	399	416	366.1	4,425.6
1988	106	550	423	320.7	4,309.9
1989	116	475	501	305.2	4,200.3
1990	137	451	506	291.0	5,054.3

Source: Gould et al. 1991; American Petroleum Institute 1993.
[a] Millions of barrels.
[b] Trillions of cubic feet.

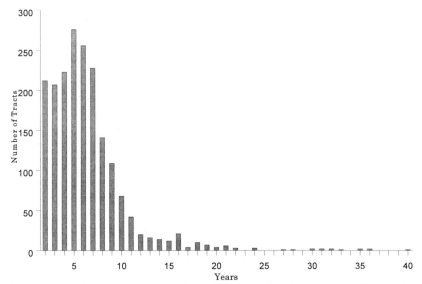

Fig. 3.1. Years between Lease Sale and Production for Leases with Production, Gulf. *Source*: Minerals Management Service 1993.

the Gulf, the average time between lease sale and production over the entire history of development in the Gulf is about five years (see Fig. 3.1, which plots the lag between finalization of the lease sale and first production on those leases that had production by the spring of 1993 (Minerals Management Service 1993)).

This relatively short average time between sales and production (considering the logistics involved), coupled with the number of leases being sold, became a driving developmental force in the Gulf and created a growing demand for improved technology (in order to explore and develop more rapidly and in ever deeper waters), greater numbers of exploratory rigs, more efficient ways to bring development platforms on line, pipelines to transport offshore production, and a massive support sector to support these other demands. The economic forces at work brought about integrated changes in the technology, infrastructure base, and physical environment and concomitant changes in the social and economic environments of the coastal Gulf of Mexico. Together these factors worked to produce a massive offshore-onshore system in a remarkably short period of time.

Nor was all the Gulf of Mexico development exclusively for the Gulf. Because of the mobility, noted previously, early offshore efforts in other regions of the world were supported from the Gulf of Mexico. Of the four production platforms in place in the Pacific at the time of the Santa Barbara spill, two of the steel jackets and two of the decks were built in Morgan City, Louisiana (Gould et al. 1991). In a similar fashion not only was early exploration in the North Sea supported by vessels and equipment from the Gulf of Mexico, but the evolution of altered forms of work scheduling meant that many of the experienced personnel from the Gulf were involved in the expansion of offshore activities around the world, a trend that continues to the present.

The Social Environment

In addition to the evolution of technological systems, the movement offshore required support from a number of social and economic systems, which in some cases had to develop new ways of operating, new networks of supply and support, and new ways of adjusting to the impacts of the growing offshore sector. While certain occupations, primarily associated with transportation or military endeavors, have historically required individuals to travel considerable distance in association with their employment, the primary reason for this pattern was the lack of a fixed geographic location around which settlement could occur. The movement into the marsh and estuaries of the Louisiana deltaic plain provided a situation that had not really existed before. Here the development was large scale, intense, and relatively permanent but was happening in environments where human settlement was not really possible. At the same time, the isolation of the activity and the time and expense to commute to it meant that commuting would have to be limited, since a point of diminishing returns is quickly reached in terms of the distance one can reasonably live from work. It is precisely this consideration that has been responsible for the concept of "labor force" being defined geographically, as an area in which people can commute to work (Killian and Tolbert 1993). If, however, time at work and time away from work can be concentrated, then space between work and residence can be increased since trips must be made less frequently.

Extended presence at the place of employment has also occurred historically, again generally involving transportation (particularly maritime) and military activities. As perhaps the most notable example, sailors commonly live and work aboard ships for months or even years at a time. Like ships, drilling operations in the marine environment required twenty-four-hour manned operations in order to be cost efficient, but three major factors worked against the "ship model" of development for marine petroleum activities as they emerged in the Louisiana deltaic plain. First, the movement was *from* traditional land-based activities *out* toward the marine environment. As noted previously, early marine operations were supported by fishing vessels, simply because the maritime expertise did not exist in the oil industry. Thus, while similar types of scheduling, though less uniform in nature, had been in effect in maritime operations since probably before recorded history, the scheduling of operations in the marsh and later offshore did not directly come out of these experiences. Second, ships (at least when they are in isolated situations at sea) are designed to be virtually self-sufficient, needing no supplies or additional personnel while en route. This negates the possibility of exchange of personnel between ports, except in emergencies. In contrast, drilling locations must be constantly supplied not only with a variety of products but also with specialized personnel to respond to various stages and potential scenarios throughout the drilling procedure. This means that constant transportation avenues must be maintained, and thus regular exchange of personnel is possible. Finally, ships are designed for maximum personnel efficiency in order to utilize as few people as possible. Marine drilling operations require many more activities than does the piloting of a ship and as a result are both more labor intensive and require a more diverse labor force.

In response to these unique variables, a form of concentrated work scheduling evolved around marine petroleum activities in the Gulf of Mexico. The common pattern for offshore work is for the employee to meet at a prearranged site to "go offshore" either by boat or more recently by helicopter. The stay offshore varies but is typically for seven, fourteen, or twenty-one days. Following the stay offshore, the employee is returned to the meeting site and has a period of time off, typically the same length as the stay offshore. When offshore, the crew is divided into two shifts, which are on for twelve hours and off for twelve hours alternatively. In effect there are four complete crews for each operating rig, two of which are offshore at any given time. More specialized personnel, who are not part of the crew necessary for the basic operation of the drilling rig (downhole logging, etc.), may work other schedules, such as on-call twenty-four hours a day. Offshore employees work, eat, sleep, and, in short, live at the place of employment. This concentration of work scheduling, which became the model for offshore development, was one of the necessary elements of social engineering for development in the isolated environs of the continental shelf. This pattern of employment has been operational in the Gulf of Mexico for over four decades (Gramling and Brabant 1986a). The manifest function was to minimize transportation time and cost, but over time two related latent functions have emerged.

First, because of the temporal scheduling of offshore work, employees can realistically seek and obtain employment considerable distance from where they live. If, for example, one works a fourteen-on and fourteen-off shift, the round trip between place of residence and place of work must only be made once every twenty-eight days. This allows considerably more time and money to be spent in transit before a point of diminishing returns is reached.

Gramling (1980a) conducted a labor survey in the eastern part of St. Mary Parish, one of the most heavily involved parishes in Louisiana in the support of offshore activities, during the height of the offshore boom in 1980.[6] Because concentrated work scheduling allowed long-distance commuting, some of these high paying jobs were in communities far from where individuals were employed. Table 3.3 shows distance traveled to work as related to the number of years that individuals had lived in their community of residence at the time of the survey. Not only does the table show that a fairly large portion of the sample commuted from over 100 miles away but that over 70 percent (120 out of 168) of these long-range commuters had been doing so, while continuing to live in communities located far from offshore staging areas, for over ten years

Table 3.3. Miles Traveled by Years of Residence, for Shift Workers

	YEARS OF RESIDENCE			
MILES TRAVELED	0–1 YEARS	2–5 YEARS	6–10 YEARS	10+ YEARS
0–25 miles	10	4	5	18
26–50 miles		1	1	5
51–99 miles	10	22	6	25
over 100 miles	38	64	34	120
Total	58	91	46	168

when they were surveyed in 1980. In fact, the sample contained individuals living in eighteen states and commuting to work offshore in Louisiana from such improbable locations as New York and California.

Further analyzing this data, Gramling and Brabant (1986b) noted that less than one third (30%) of the 381 offshore workers in their sample lived within 100 miles of where they met to go offshore. Over eleven percent (11.5%) of the employees lived more than 500 miles from where they worked. Similarly, Centaur Associates (1986) found that of the estimated 21,847 employees working offshore in waters off the state of Louisiana in 1984, 2,167 (9.9%) did not live in the state of Louisiana, and the overwhelmingly majority did not live in the parish (county) where they were employed.

Second, this nontraditional scheduling of work allows decisions concerning the site selection of drilling activities to be made with less regard to human settlement patterns, since decision makers can count on long-distance commuters. These two points are obviously different sides of the same coin: the first refers to decisions by employees, the second, to decisions by employers. However, taking these two points together and increasing the length of the period of intensive work and correspondingly the period of time off, a particular economic activity, such as an offshore drilling rig, can be sited literally in any coastal region of the world and the labor force to man the platform can be from any other region of the world. Initial development off the north slope of Alaska was staffed largely by experienced offshore drilling crews from Louisiana and Texas. The scheduling of work was simply changed to thirty days on and thirty days off to allow feasible commuting. Likewise, as offshore exploration and development moved off the North American continent (to the North Sea, the Persian Gulf, Indonesia, etc.), work scheduling was extended to ninety days, and the same individuals who had obtained their experience in the northern Gulf of Mexico were employed in these locations.

In fact, in the North Sea, where U.S. companies dominated the early offshore activities, it quickly became obvious that if left to their own decisions, the corporations would employ the majority of their overseas labor from the experienced labor force located near the Gulf of Mexico. The corporations involved profited because they were relieved of the necessity to train or relocate employees. Thus, few of the employment benefits associated with even massive offshore development were accruing to the countries contiguous to that development. Once this trend became clear, the countries involved, led by Norway, moved legislatively to require that quotas of local residents be employed and local supplies be purchased in virtually all offshore activities. This led to an economic development model, which Noreng (1980) calls the North Sea Model, and House (1984) refers to as regulated capitalism.

The point is that this particular altered form of work scheduling produced such a successful combination of a geographically separated economic activity and a labor force that legislative intervention was required to alter it. For certain highly skilled individuals in the oil industry this pattern has become a way of life. Much like experienced construction foremen or consultants, these individuals work wherever the action is. Unlike construction foremen, they continue to "live" in a traditional place of residence. Unlike consultants, their employment is regular and frequently with the same corporation and in the same location for years. Thus, the offshore industry and certain highly skilled individuals have in select economic sectors literally created a worldwide labor market and eliminated the necessity for the geographic concentration of human settlement in the vicinity of economic activity.

While these operations were supported by maritime operations, they were not a part of these operations, and neither the oil worker in the marshes in the 1930s and 1940s nor the vessel operators and crews that got them to work were a part of the long tradition of labor relations that had built up around seafaring (Forsyth 1989;

Forsyth, Bankston, and Jones 1984; Forsyth and Gramling 1987). While this may seem to be a minor point, it is not. The emerging occupational structure of work associated with oil and gas extraction in the marine environment both diverted from and shared with elements of the more traditional maritime industry. These shared elements and diversions have shaped labor relations and the legal environment of offshore work to the present. The gradual extension of land-based oil extraction activities in the 1930s and 1940s into the marine environment, the support of those activities by an emerging class of vessels, which were too large to be considered "boats" and too small to be "ships," and the isolation of this system in inaccessible areas meant that one of the emerging industrial issues of the day, unionization, was left behind.

The late 1930s and early 1940s saw the growth of the union movement in the United States, following the passage of the National Labor Relations Act of 1935, and although the earliest maritime unions can be traced to the late 1800s, the primary thrust for the seaman unions paralleled the time period of the union movement in general (Forsyth 1989). By this time established patterns of employment surrounding oil and gas extraction activities had developed in the coastal Gulf of Mexico, and several features of that activity ensured that, unlike seamen, these and later offshore workers in the United States never became unionized.

Shrimpton and Storey (see also Storey and Shrimpton 1986) summarize the problems with unionization at isolated mine sites in northern Canada where both the scheduling of work (Gramling 1989) and the isolation is very similar to that in offshore oil:

> The problems that "fly-in" [there are no roads, so miners fly in to work] pose for unionism centre on the facts that the entire mine site, including the mine, camp and airstrip, are under company control, and that at any one time only half the workforce is at the site (half of whom are working), while the other workers are at home in a large number of widely dispersed communities. Access to the site has to be negotiated with the employer, or achieved through government intervention, and it is difficult to hold meetings and votes, achieve group solidarity, and provide union training. In the case of a labour dispute it would be very difficult to picket, prevent the

use of non-union labour, or maintain solidarity among the dispersed workers. There is also no "community" support, in the form of the support from the spouses and children living in the single-industry town that has proven so important in numerous labour disputes. (Shrimpton and Storey 1987: 4)

Although offshore workers in Europe are largely unionized, the offshore industry in Europe (North Sea) developed in the 1960s around relatively mature technology and was organized along the lines of the existing shipping industry and the accompanying maritime unions. Maritime unions have been successful, at least partly, because of licensing requirements for every entry into critical jobs and the comparative labor scarcity created by these entrance requirements.

In addition, ships (because of their draft and the facilities needed to load and unload them) come into a limited number of ports. Union halls were established in the 1930s in all of these major ports, which provided a gathering point for seamen and a focus for union activities. There is no comparative point of concentration or licensing requirement for those involved in offshore oil and gas activities in the United States. As noted earlier, because of the smaller scale and limited requirements, early marsh and offshore activities were spread out over numerous local waterways across the Mississippi deltaic plain. Thus, in the United States where the offshore industry evolved as a gradual extension of onshore practices, the industry has successfully avoided unionization throughout its sixty-year history.

Ironically, when the issue of representation by a seaman union for offshore oil workers did come up, it was fought by a competing union. The question arose when the Sailors Union of California sought an election to represent Richfield Oil Company workers offshore of California. The request was contested by the Oil Chemical and Atomic Workers of America Union, which represented Richfield's onshore workers, and the protest prevailed with the National Labor Relations Board (*Drilling* 1958b). The offshore workers were not, however, later represented by the Oil Chemical and Atomic Workers of America Union.

What the industry did not avoid that was a direct result of the gains of the maritime union was protection for injured workers under the Jones Act, a near fluke of the U.S. judicial system that has led to a plethora of personal injury suits in the Gulf of Mexico (Beer 1986). Following the sinking of the Titanic in 1912 and the outbreak of World War I in 1914, national attention focused on the protection of sea lanes and on seamen who manned American vessels (Beer 1986). In 1920 Congress passed the Merchant Marine Act (section 33 of which has come to be named the Jones Act), which provided a recourse (including a trial by jury) for "seamen" for the collection of damages for injury suffered "in the course of his employment" (46 U.S.C. 688(a)). Congress recognized the seaman as a special type of plaintiff for two reasons: the seaman's *exposure* to the perils of the sea and the relative *isolation* from recourse and remedies. The seaman performs dangerous acts under orders and cannot quit and simply walk away when a vessel is at sea (Caffery 1990). As a consequence the Jones Act makes it easier for seamen to establish employer negligence and to claim damages than with other occupations. Neither "seaman" nor "vessel" was defined by the act.

In 1955 a suit was brought against the Texas Company in U.S. District Court by the estate of a worker who was killed on a submersible drilling barge, and the court returned a settlement under the provisions of the Jones Act. Although the Fifth Circuit Court of Appeals overturned the decision, the Supreme Court upheld the original decision. The effect of this was to define submersible drilling barges (and later offshore rigs) as "vessels," and employees on these vessels as "seamen" entitled to protection under the Jones Act. Because work in the oil field in general and offshore in particular was, and is, dangerous, this decision touched off a wave of Jones Act litigation in coastal Louisiana and Texas, which continues to the present (Beer 1986).

Working Offshore

The geographic job mobility noted earlier constitutes one of the better examples of both positive and negative impacts distributed by offshore development. One of the most positive impacts of the concentrated work scheduling associated with working offshore is that it allows individuals living in depressed economic areas,

sometimes far from the coast, to maintain a high paying job offshore while continuing to live in (often rural) communities with
their family and social networks (Gramling 1989).

There are, however, a variety of individual problems, which occur with offshore employment, such as lack of privacy on the job,
loneliness, limited recreational facilities, and perceived risk of the
job. Also, many offshore jobs are, quite simply, difficult because
they involve arduous labor in the midst of hostile elements. In addition, time off the job may also be unrewarding. Extended periods of leisure time, when all of one's friends are working, are less
than ideal. Boredom probably leads to more second jobs by offshore workers than does financial need.

Individuals generally compartmentalize their lives in many
ways, but a common one is to divide time between work and leisure. Shift work interrupts this normal compartmentalization and
interrupts the individual's interaction within a variety of social
institutions (Laska et al. 1993; Gramling and Forsyth 1987). There
have been some attempts by local institutions to adjust to this in-
terruption. For example, the University of Southwestern Louisiana
at Lafayette, near the center of offshore production, offers classes
that meet every other week for those employed offshore. However, the most difficult adjustment for offshore workers is with
their family life.

Gramling (1984) noted some specific family problems related to
decision making, child rearing, and sex, which are associated with
offshore work, but Gramling and Forsyth (1987) in a more comprehensive approach examined the general relationship between
work scheduling and family interaction. Using the construction of
reality approach (Berger and Luckmann, 1966; Berger and Kellner,
1964), the family was conceptualized as a small social system,
which in modern industrialized cultures is less and less guided by
traditional norms and increasingly must operate by a set of rules
that are largely negotiated, by the members of the family. These
continually updated rules or constructs define appropriate behavior for each individual family member, as well as what constitutes
appropriate interaction among family members.

Successfully arriving at a consensual definition or an agreed
upon set of ground rules is an ongoing enterprise as each new

situation that the family faces must be defined and rules worked out for how members should act in these situations. "Problems" arise in families when they are unable to arrive at a set of agreed upon rules. The process is essentially one of negotiation, requiring human interaction. Thus, for a viable definition or set of rules to emerge, the presence of those to whom the definition or rules will apply is necessary. Offshore work interrupts the presence of one member of the family, usually the male, as offshore workers are overwhelmingly male. These prolonged and periodic absences mean that inevitably new situations and problems will arise and be made accountable while one family member is not present. Because of the situational nature of these "solutions," the absent member may not understand why and how decisions were made and may thus disagree with them. If these absences continue it may become impossible to maintain a single coherent set of rules for the social system of a specific family.

A common occurrence with the families of offshore workers is for two discrete social systems with two sets of rules to evolve. One set of rules defines areas of competence, lines of authority, and guidelines for interaction when the offshore worker is absent; the other is in operation when he is home. The relationship between these two situations seem to fall into a limited number of adaptations (Forsyth and Gramling 1987 identify five), ranging from quasi-replacement of the husband/father in the decision-making process during his absence (usually by a male kin) to the acquisition of the "periodic guest" role by the husband/father when he is home with little input into the day-to-day operation of the family. There are also almost inevitably areas where the two constructs conflict. This may not be directly problematic for the offshore worker, as he is only familiar with one set of rules, but the situation can be very problematic for remaining family members as they must literally live and interact within two alternating social systems with two sets of normatively defined behavior (see Forsyth and Gramling 1987; Forsyth and Gauthier 1991, 1993). In short, the structure of offshore employment creates a situation where, for at least a majority of the members of an individual family, the potential for anomie[7] with all the attendant problems is high (Durkheim 1951).

Economic Adaptations to Mobility

While the development offshore, starting in the 1950s, generated tremendous economic growth along the northern Gulf of Mexico, the interaction of the mobility of that activity with the structure of ownership by the major multinationals produced an unusual economic environment in the Gulf. The major multinationals, because of their immense capital potential, own (or through the leasing process acquire control of) the potential oil and gas resources in the reservoirs and the refineries required to produce a marketable product. In addition, as the *Exxon Valdez* reminds us, they may own the means of delivering raw materials to the refineries and the distribution system allowing the marketing of finished products from the wholesaler to the Texaco station on the corner. What these fully integrated corporations most notably do *not* own are the corporations and equipment that address the various pieces of the process necessary to get the resource in the ground to a point where it is available to be transported to a refinery. Although the necessary capital for the first offshore operations was only available from the largest oil companies, the concept of equipment leasing (including offshore drilling rigs and support vessels) and independent offshore supply companies (e.g., to supply drilling mud, water, etc.) quickly emerged as offshore operations began to standardize in the 1950s (Lietz 1956; *Drilling* 1956c). It is here in this most temporary of situations where the opportunity for the myriad smaller players in the offshore game exists and where much of the expertise in the Gulf infrastructure remains (see epilogue).

For example, let us say British Petroleum is the successful bidder on an offshore tract. British Petroleum will contract with an offshore exploration company to do the seismic exploration, and if the results warrant, with perhaps another company to do the necessary exploration for underwater hazards. Then they will contract with an offshore drilling company to drill an exploratory well. The drilling company owns offshore rigs but contracts with a host of other specialized companies to support the drilling operation, providing support vessels, catering, potable water, drilling mud, casing and drill string, and even crews. Many of these companies further contract with still another level of support companies to provide additional services. Thus, a corporation that owns and leases offshore supply boats will probably engage another company to provide offshore safety classes for their employees.

If the exploratory well finds commercially feasible quantities of oil or gas, then the oil company will contract with a fabrication company to produce an offshore production jacket and perhaps another company to produce the deck. Additional companies may move the completed jacket and install it, requiring support (e.g., a diving company) throughout the process, and install the completed deck structure on the jacket. At this point, perhaps, another drilling company gets involved drilling the production wells, also being supported by many firms through a process that may last several years. After completion, a pipeline company connects the platform to the existing pipeline network, also being supported with crews, fuel, water, pipe, coating, divers, and so on throughout. If something breaks down during any of these operations or later modifications need to be made, a host of additional corporations swing into action with specialized services and tools. Once the resource is available to be delivered, the oil company again assumes primary control and profit.

Several attributes of this process are noteworthy. First, it is in this niche between initial ownership and availability for transportation (exploration-development) that the primary opportunity for local entrepreneurial activity exists. Second, this is one of the more capital intensive aspects of the enterprise, and products that facilitate the process are very valuable. Third, this part of the process from acquisition of lease to availability for refining is, almost by definition, the most temporary. Finally, because of the uncertainty involved at each stage of the exploration-development process and the necessity to rely on services outside one's control, many jobs cannot be bid at a fixed cost, as say the construction of a house or road can be. A supply vessel company cannot bid a fixed price for supplying a drilling rig, because those responsible do not know how long the operation will take. Often they cannot even give a daily rate in advance because they do not know what difficulties will be encountered (up to and including such factors as the weather) and, as a consequence, what vessels will be required and for how long. Personal services are similarly constrained, as are the production of individual specialized products used in the system. Thus, the whole support sector works on an hourly or at best a daily basis and on a high markup for specialized products.

Given these attributes and an aggressive federal leasing program, a very unrealistic entrepreneurial environment arose surrounding offshore activity in the northern Gulf of Mexico. Because payment was on an hourly or a daily basis and because of the large markups, the emphasis came to be on speed not costs. When a $60,000+ a day offshore drilling rig is down for lack of a part, the question is not how much but how soon. Similar metrics are at work to a lesser scale with a $6,000 a day supply vessel or with skilled labor when the demand for a product is high. The fabrication yards in Morgan City were paying $18 an hour for welders in the mid-1970s and could not get enough. Thus, while the economic environment provided jobs, entrepreneurial opportunities, and in many cases fortunes, it was only during one phase of the operation, a phase that only occurred for a limited time. The extreme adaptation to this narrow niche was to prove to be problematic when that time passed.

By the mid-1960s, the offshore system was a mature industrial complex, fully integrated into coastal Louisiana and Texas. Operations were underway in the Pacific off southern California and had started in the North Sea. In less than two decades, the industry had evolved into a system capable of working throughout the world. During the next decade, while this worldwide expansion continued, operations in the Gulf grew at a steady pace. Lease sales proceeded on an annual basis in the Gulf, and the steady annual average increase of over 100 platforms continued to drive the development of the support and construction sectors of the economy. Although the January 28, 1969, spill in the Santa Barbara Channel halted additional Outer Continental Shelf lease sales in the Pacific and postponed initial sales in the Atlantic and Alaska, the spill did not affect the Gulf. For the next five years, while the Outer Continental Shelf of the majority of the United States was closed to leasing, the leasing program in the Gulf and the accompanying developmental pressures moved steadily ahead.[8] During the same period several national and international events were unfolding that were to directly affect the offshore leasing program. These events are discussed in chapter 4.

Chapter 4

Political Storm Clouds

OPEC, CALIFORNIA, AND THE EMBARGO

THE FORMATION OF OPEC

Even though enormous forces to encourage consumption had been put into place following World War II (see chapter 1), exploration and production technology moved ahead even faster. Production on the Gulf of Mexico Outer Continental Shelf was in full swing by the late 1950s, discoveries were still being made in coastal Louisiana and Texas, and the Middle East was found to be floating on oil. With reservoirs close to the surface and close to the sea, which facilitated transportation by tanker, production costs in the Middle East ranged from five to ten cents a barrel. Sold at Gulf of Mexico prices under the Achnacarry agreement, the profit provided a powerful incentive to encourage production. In addition, Soviet oil was beginning to come on the world market.

The transportation structure of the market was also changing. Following World War II and the Marshall Plan, which encouraged the transition from coal to oil, much of Europe was supplied with Middle Eastern oil via the Suez Canal. When Nasser closed the canal in 1956 in response to the British and French invasion to retake it, the progenitor of the super tanker was born. The industry moved quickly to build 250,000-ton tankers. These ultimately proved to be so cost effective that eventually it became cheaper to bring oil from the Middle East to Europe around Africa. As a result of all of these factors, by the late 1950s concerns about oversupply once again came into the forefront.

Following a series of unsuccessful attempts at voluntary controls, in 1959, a reluctant President Eisenhower announced a quota on foreign oil. The quota resulted in a decline in the demand for Middle Eastern oil, and the following summer Exxon abruptly announced a 14 cent per barrel cut in its posted price for Persian Gulf oil (Atkins 1973). The result of this simple profit-motivated action was to have far-reaching consequences. By September of 1960 the Ministers of Oil of Iran, Iraq, Kuwait, and Saudi Arabia had met in Baghdad, and the Organization of Petroleum Exporting Countries (OPEC) was born (Ghanem 1986). OPEC had little initial success in establishing quotas, and, as a result, the world surplus of oil continued. Although the major oil companies did not take the organization seriously, they did consult with the new Kennedy administration in order to be certain that they would have the permission of the antitrust division of the Justice Department to act jointly if OPEC became a problem (Senate Subcommittee on Multinational Corporations 1974). Failure to recognize the potential of OPEC was to have serious consequences, which eventually transformed the Outer Continental Shelf leasing program in the United States.

OFFSHORE CALIFORNIA

During this same period events were unfolding on the West Coast that were also to eventually alter the Outer Continental Shelf program (Cicin-Sain et al. 1992). By 1957 the Monterey-Texas Company had producing wells in the Pacific from an artificial island off Seal Beach (*World Oil* 1957d). In 1958, borrowing from Gulf of Mexico technology,[1] the first true offshore platform in the Pacific (Hazel) was set at 100-foot depth in state waters, offshore from Summerland. By 1966 the state of California had leased all state offshore lands from Point Conception on the west eastward to the Ventura County line, practically the entire coastline of the Santa Barbara Channel. The exception to this was a sixteen-mile sanctuary, adjacent to Santa Barbara and the Golita valley, where the strongest protests against offshore development forced the state's hand (Steinhart and Steinhart 1972). To date there have been over 3,500 development wells drilled in state waters off California (American Petroleum Institute 1993).

Spurred by this growing movement offshore into state water, the first federal lease sale in the Pacific was held in May of 1963 (Gould et al. 1991). Although 41 tracts off central and northern California were leased in this first sale and 101 in a sale off the Washington and Oregon coasts the following year, exploratory drilling found no commercially feasible quantities of oil. It was not until 1966 that tracts were offered in the promising Santa Barbara Channel area. In 1966 a one block "drainage" sale was held, over local protests, to counteract the effects of production on state leases adjacent to federal waters, and industry response foretold of the perceived potential of the Pacific Outer Continental Shelf. A consortium of three of the major oil companies[2] bid $21,189,000 for a 1,995.48-acre tract, or $10,618.50 an acre (*Offshore* 1967a).

With its movement into the Santa Barbara area and with the next lease sale in 1968, the Department of Interior, for the first time, began to face real opposition to its offshore leasing program. The proposal for sale P4 (originally scheduled for 1967) brought forth a storm of protests within Santa Barbara County and a county report opposing the sale (Steinhart and Steinhart 1972). As a result of the protests Interior delayed the sale and agreed to a two-mile buffer around the Santa Barbara sanctuary. The industry was so anxious to proceed in the Pacific that five mobile drilling rigs lay idle in California ports during the delay (*Offshore* 1967b).

The delayed February 6, 1968, sale turned out to be a record one from a financial perspective with total bonus for the leased blocks of over $602 million, surpassing all previous Gulf or Pacific sales (Gould et al. 1991). Within weeks of the sale, two drilling rigs were on location and working (Armstrong 1968). Slightly less than a year later, January 28, 1969, on one of the tracts sold (tract 402), the fifth well being drilled on Union Oil's Platform A blew out around the casing below the sea floor. The strata below Platform A continued to leak throughout much of 1969, although the worst was over by mid-March. Estimates of the total spill range from one to three million gallons.

More important than the exact amount of oil spilled is the fact that the Santa Barbara oil spill became a national media event throughout the spring of 1969. Scenes of helpless birds mired in a layer of oil became common and, coupled with the inability of either the oil companies or the federal government to stop the flow, had far-reaching effects (Molotch 1970). The spill has been variously credited with lending support to the emerging environmental movement (see Dunlap and Mertig 1992 for a complete analysis), Earth Day in 1970, and the passage of the National Environmental Policy Act (NEPA) in the waning days of 1969. As a result of the spill, lease sales were suspended in the Pacific and proposed sales in the Atlantic and off Alaska were postponed. The immediate effect of the suspension was to halt the planned development curve of the Outer Continental Shelf outside the Gulf of Mexico and consequently to enhance the importance of Middle Eastern oil.

The 1966 and 1968 sales in the Pacific put the Department of Interior into a confrontational position with the local residents and the residual aftershocks of the Santa Barbara spill later brought the state of California, together with other coastal states on the East and West coasts and ultimately Congress, into the fray positioned against Interior's offshore leasing and development plans. The end result of this posturing, which began in the 1960s, was that by the early 1990s, virtually the entire Outer Continental Shelf of the lower forty-eight states with the exception of the Gulf of Mexico was closed to the very agency within Interior whose mission it was to lease the Outer Continental Shelf (see chapter 6).

BACKGROUND TO THE EMBARGO

Another critical, less publicized, event that was to ultimately have far-reaching implications for the offshore leasing program also happened in 1969—the rise of Muammar el-Qaddafi to power in Libya, through a military coup (see Anderson and Boyd 1983 for details of this period). By 1969 Libya had risen from the role of a poor desert country, subsisting largely on rent from British and U.S. military bases, to a producer of 3 million barrels of oil a day. Following the closing of the Suez Canal during the 1967 Arab-Israeli war, Libya, since it was located on the Mediterranean, became the centerpiece of a strategy to provide an alternative to Middle Eastern oil for Europe. A key player in this strategy was the Occidental Petroleum Company, which had major holdings in Libya and markets in Europe.

Qaddafi was an admirer of the Pan-Arab position of Abdel Nasser and like Nasser, who drove the British out of Egypt, quickly moved to stem foreign and multinational influence. By the end of 1969, Qaddafi had announced what essentially amounted to the expulsion of the British and U.S. air bases, and the scene was set for confrontation with the oil companies.

In January of 1970 Qaddafi demanded an unprecedented 40 percent price increase for Libyan oil and a larger share of the profit to Libya, and he threatened to stop production if his demands were not met. Several factors were in Libya's favor. First, Nigeria was embroiled in a civil war, which took its oil out of production. Second, an unusually cold winter in Europe had drawn down supplies. Third, the building of tankers for the trip from the Middle East by going around Africa had not yet caught up with the transportation bottleneck caused by the second closure of the Suez Canal, as a result of the 1967 Arab-Israeli war. This put Libyan oil, which could be delivered across the Mediterranean, at a premium and Occidental Petroleum, whose primary market was Libyan oil for Europe, in a bind. Realizing this, Qaddafi began to cut Occidental's production. By early September after three cutbacks and lack of support from the other companies with Libyan interests, Occidental capitulated. With the capitulation of Occidental, facing new eminent threats to totally cut off their production in Libya, and a complete lack of support for resistance to Qaddafi by the Nixon administration, most of the major buyers of Libyan oil folded and met Qaddafi's demands before the end of the month.

The floodgates were open. With Qaddafi's success, the members of OPEC had a new model for negotiations with the multinationals, and they lost no time in applying it. Demands similar to the gains secured by Libya were soon made to the oil companies, with similar threats of shutting down production or even nationalization of holdings if demands were not met. Initially the oil companies closed ranks against the demands and, with the blessings of the Nixon administration, which agreed for all practical purposes to set aside antitrust considerations, demanded to negotiate with the OPEC countries as a block. On the eve of the negotiations Kissinger began to waver, and subsequently the Nixon administration withdrew its support for the oil companies position. By the middle of February, conditions similar to those won by Libya were in effect for the Persian Gulf members of OPEC. By the end of March, Qaddafi had made new demands, which the companies met (Anderson and Boyd 1983).

With the events in Libya and the Middle East during 1969–1970, control of the world oil production passed from the cartel of multinational corporations to the producer states. Between 1971 and 1973 the price of Middle Eastern crude doubled. A working relationship between the oil companies and OPEC quickly developed. The oil companies passed the increased cost of oil, along with increased profit margins, to consumers, and the early 1970s were banner profit years for the multinationals.

There were, however, warning signs on the horizon. In 1970 Nixon signed the Clean Air Act, which began to move the country away from coal and toward cleaner fuels, particularly natural gas. Gas, however, continued to be federally regulated at what many thought was an artificially low price, and whether or not this was the case, there had been no big push by industry to increase available supplies. In addition, the projected development curve of the Outer Continental Shelf, outside the Gulf of Mexico, had been severed by the moratoria following the Santa Barbara spill. Finally, to the extent that long-range planning was being undertaken, the projected 1.7 million barrels a day, which would ultimately flow through the Trans-Alaska Pipeline, had been delayed by heated opposition and the failure of the partners in the venture to come up with a credible environmental impact statement (Gramling and Freudenburg 1992b).

Sitting comfortably in retrospect, Anderson and Boyd sum up the position of the Nixon administration in the spring of 1971.

> What could hardly be missed was that a dozen significant energy decisions, domestic and foreign, had been made by the Nixon administration in its first two years, all with little regard to their energy security implications, each responding basically to a nonenergy stimulus, each skewed by a nonenergy goal, none coordinated with the other. All, however,

were contributing to an eerily uniform result: the generating of a several-million-barrel-a-day oil shortfall that must with blindsiding effect converge upon the oil market in two to three years, unless there was a redirection of policy *now*. (emphasis in original) (Anderson and Boyd 1983: 289)

THE EMBARGO

Throughout 1971 and 1972, the producer nations moved to solidify their position through "participation" (increasing nationalization of corporate holdings) with the multinationals. As part of these agreements, the producer nations were transporting and refining increasing amounts of their oil. This forced the companies to curtail some of their long-term supply contracts to utilities, independent refineries, and other large users and, in turn, forced these buyers into the more volatile spot market. Also throughout this period, it was becoming increasingly clear, particularly to the oil companies, that their Middle Eastern supply of oil was being threatened by U.S. policy in the area with its support of Israel and Israel's continuing occupation of Arab territory (Anderson and Boyd 1983).

On October 6, 1973, Egypt and Syria invaded their occupied territories, and their initial success led Israel to call on the United States for support. Representatives of the oil industry clearly warned the White House that support for Israel would bring an embargo (Sampson 1975), but bogged down in Watergate, Nixon used the publicity opportunity to call on Congress for aid to Israel. Soon after this the Arab members of OPEC initiated an embargo against the United States. Although the embargo was over by the spring of 1974, its effects continue to the present. When the volatile spot market reacted and demonstrated to the OPEC members that the market would bear much more than the current posted price of oil, OPEC raised the price. In a three-month period, from October of 1973 to January of 1974, Middle Eastern oil prices soared from $5.12 a barrel to an official OPEC price of $11.65 a barrel and at times much higher on a nervous spot market. This came on top of the doubling of the price between 1971 and 1973 (Darmstadter and Landsberg 1976). These increases were only the beginning of a steady climb that would see crude oil peak at over $30.00 a barrel seven years later.

The impact of this trend on the Outer Continental Shelf program was twofold. First, and quite directly, in the Gulf, where leasing had continued after the Santa Barbara oil spill, the rising prices within a year drove the bids on leases and exploration and development activities up (see Table 3.1; Gould et al. 1991). A more circuitous impact came from the Nixon administration's response to the "sudden" crisis.

By 1973 the Nixon administration was awake to the situation and President Nixon proposed a quick fix to the shortages and price rises following the embargo by offering the American public Project Independence, which was supposed to bring energy independence by 1980. In addition to a renewed call for the Trans-Alaska Pipeline, the plan instructed the Secretary of Interior to increase the Outer Continental Shelf acreage offered for lease to 10 million acres for 1975 (about three times what had been planned) and to begin lease sales in "frontier" areas, those that had never seen leasing and those that were closed following the Santa Barbara spill. A Senate staff report (Magnuson and Hollings 1975) noted a number of problems with the plan. First, there was the overly optimistic estimates of the reserves on the Outer Continental Shelf and hence the questionable ability of the Outer Continental Shelf to meet the nation's energy needs. Second, the report noted that the increased offerings would probably result in lower dollars per acre for bonus bids for Outer Continental Shelf leases given the "law" of supply and demand. Third, there was also the possibility for increased impacts on coastal areas as a result of the increased activities. And finally, the industrial capacity to explore the increased acreage (drilling rigs, support vessels, etc.) simply did not exist, a fact also recognized by the industry (*Offshore* 1974). Although Nixon's tenure in office was terminated by his resignation, President Ford continued his Outer Continental Shelf policy. Thus, in spite of the noted problems, increased leasing moved ahead, though lease sales never approached the 10 million–acre goal. In 1975 sales resumed on the Pacific Outer Continental Shelf and in 1976 on the Atlantic Outer Continental Shelf.[3] The first sale also went forth on the Alaska Outer Continental Shelf in 1976. In the Gulf, acreage sold in 1975 exceeded that of any previous year (Gould et al. 1991).

Outside the Gulf, Outer Continental Shelf leasing had become an increasingly unpopular activity. The initial sales in Alaska and off the East Coast and the resumption of sales in California in the mid-1970s met with resistance from various interest groups and state governments. In California, the Santa Barbara spill had galvanized local groups and increased pressure statewide for protection of the coast. The fledgling environmental movement, which gained support from the media coverage of the Santa Barbara oil spill, began to be a factor in the opposition for further Outer Continental Shelf development and has remained a factor to the present.

In 1972 California passed Proposition 20, which created the California Coastal Commission with authority over almost all types of development on the coast and out to the three-mile limit of state waters. The commission's first task was to develop a plan detailing acceptable uses of the coastal zone. The plan went before the state legislature in 1976 and, after a bitter fight, passed. While the plan had no direct control over Outer Continental Shelf lands, the process set up by the Coastal Commission provided access to a variety of information, rallying points, and communication vehicles for various users of the coast. In short, it made organization to oppose Outer Continental Shelf activities easier (Kaplan 1982), and opposition began to reemerge shortly after the announcement of the proposed resumption of leasing. The initial sale in California after the moratorium (sale 35), proposed for 1974, was opposed by the state, which sued, maintaining that the requirements of NEPA had not been met. The state ultimately lost in court, but in the meantime two sales proposed for 1976 and 1978 were dropped. The next sale (48), originally proposed for 1977, was delayed until 1979, after a suit brought by the county of Santa Barbara failed to halt it. During the same period, not coincidentally, proposals to establish marine sanctuaries (excluding Outer Continental Shelf oil and gas activities) surrounding the Santa Barbara and Farallon islands went forth to the National Oceanic and Atmospheric Administration, under the 1972 Marine Protection, Research and Sanctuaries Act.

Sale 48 raised an issue that to date has not been fought to its legal conclusion—the relationship between Outer Continental Shelf leasing and federal consistency under the 1972 Coastal Zone Management Act (CZMA). Under this voluntary program, states with federally approved coastal zone management programs would receive federal funds for planning and management in their coastal zones. A clause in the CZMA required that federal activities be consistent with the requirements of the state programs, a logical requirement, since it hardly made sense to use federal funds to put management plans in effect if an agency of the federal government could arbitrarily act to override those plans. Thus, the consistency clause in the CZMA requires that the federal agency certify that its activity is consistent with state guidelines and submit that certification to the state for verification. While there was general agreement that Outer Continental Shelf exploration and development fell under the activities requiring consistency, with sale 48, the California Coastal Commission requested the Secretary of Interior to certify that the sale itself met consistency. Interior refused, arguing that the sale itself produced no potential impacts. California countered that all other activities flowed from the sale and that the purchase of a lease implied a right to develop, since Interior could hardly lease a tract and then deny the lessee the activity for which they had purchased the lease. In addition, the stipulation of the leases required timely development. A Justice Department review of the issue and later a Department of Commerce opinion both agreed with California, but because no tracts were actually in dispute (i.e., California would have agreed with the consistency determination), the issue was not resolved (Kaplan 1982).

The same pressure to produce offshore on the East and West coasts also affected offshore activities in the Gulf of Mexico. However, the net result of those pressures was very different in the Gulf.

Chapter 5

Boom and Bust in the Gulf

Throughout the 1960s and early 1970s, offshore exploration and production continued to increase in the Gulf of Mexico. It was during this period that offshore energy development expanded to become a worldwide phenomena. Exploration in the North Sea began in the late 1950s, and by the mid-1960s leasing and development was underway. During this same time period, exploratory drilling was proceeding in the Persian Gulf, off Africa (Nigeria), the Far East (Japan, Borneo, Australia), and Cook Inlet in Alaska.

Still the major development was in the Gulf of Mexico. The expansion in the Gulf during this period was primarily from off the Louisiana deltaic plain westward to western Louisiana and Texas and southward into the deeper waters of the Gulf. The Louisiana deltaic plain was the clear leader both in leases sold and in platforms sited until the late 1960s and early 1970s, when western Louisiana and Texas began to catch up.

CONTINUING TECHNOLOGICAL DEVELOPMENT

Driven by forecasts of growing energy consumption (*Offshore* 1966b, 1966c), throughout this period the evolution of the offshore industry continued with an emphasis on movement into deeper water and more hostile environments. The loss of the drilling rig *Sea Gem* in the North Sea in 1965 (*Offshore* 1966d) indicated the dangers associated with the new environments, and the adaptation of Gulf of Mexico technology to the more hostile conditions elsewhere, a process that was already underway, received new impetus. By 1966, the first of a new generation of massive offshore semisubmersibles and drillships began to be available for work virtually anywhere in the world, many of them constructed in the Gulf of Mexico (*Offshore* 1966e, 1966f, 1966g, 1966h) and many of them going to the North Sea (*Offshore* 1967c). The North Sea scene had actually begun in 1959 with the announcement of the Groningen gas field off Holland, the size of which was realized by 1962, starting a multimillion dollar drilling effort that was in full swing by the mid-1960s (*Offshore* 1967d). By 1967 the extent of the potential for the North Sea was beginning to be realized as new strikes were reported (*Offshore* 1967e) and the first North Sea production came ashore (*Offshore* 1967f).

This same period saw the beginning of sophisticated underwater survey techniques (Hill 1966), including side-scan sonar (*Offshore* 1966i), and technological advances in deep water diving and pipeline laying (Black 1966; Lahm 1966), which supported the direction the offshore industry was headed. By 1967 there were seventy-five countries worldwide exploring for offshore oil and gas and twenty countries successfully producing offshore (Weeks 1967).

Throughout the 1960s, however, the primary action was in the Gulf of Mexico, and Gulf technology was state-of-the-art. A number of offshore "giant" reservoirs (producing over 1,000 barrels a day and estimated to contain over 100 million barrels) had been identified in the Gulf[1] (*Offshore* 1967g). A controlled federal leasing program was leasing tracts for top dollars (Table 3.2 Chapter 3; *Offshore* 1967h; Schempf 1968a) and much optimism was being expressed by operators in the Gulf (*Offshore* 1969a).

Gulf technology had little competition. By 1967 McDermott had built, in their Bayou Bouef yard near Morgan City, two platforms for Shell oil and installed them in over 300 feet of water in the Gulf (*Offshore* 1967i). In 1968 the *Glomar Challenger*, a state-of-the-art drillship, built in Orange, Texas, began a series of world wide trips for the Scripps Institute, to take deep bottom cores for research purposes. The technology involved was inconceivable only several years before. Before the trips were over the *Glomar Challenger* would drill an almost 3,000-foot hole in over 16,000 feet of water (*Offshore* 1969b).

Construction of offshore drilling rigs began what can only be described as an incredible construction boom with most rigs continuing to be built in the Gulf. During 1966 shipyards around the world had thirty-four drilling units (17 jackups, 10 submersible/floaters, and 7 drillships) under some phase of construction (*Offshore* 1967j), a figure that climbed to forty by the following year (Schempf 1968b).

The 1969 Santa Barbara oil spill and the ensuing moratoria on offshore drilling in most areas other than the Gulf of Mexico, coupled with an aggressive rig construction program, forced many of the major oil companies to look to other areas of the world for offshore discoveries, and this reinforced an already aggressive move overseas by the drilling and support companies (*Offshore* 1968). By the mid-1970s, 75 percent of the world's offshore drilling operations were abroad, and the North Sea was the prime offshore area (Feder 1974). Through a deliberate government initiative, Norway began to surface as a leader in offshore technology, and Norwegian companies began to compete with their U.S. counterparts for a world market during the 1970s (*Offshore* 1967k; Hansen et al. 1982). As early as 1973, 34 percent of the semisubmersibles[2] under construction were for Norwegian firms, and 24 percent of the semisubmersibles under construction were in Norwegian yards (Thobe 1973).

THE BOOM

The rapid growth in Outer Continental Shelf activities in southern Louisiana and in the North Sea resulted in tremendous demand for goods and services and in increased employment opportunities. These, in turn, resulted in immigration and population growth (see Table 5.1). The growth of the population and commercial activities placed strains on existing transportation networks, community infrastructures, and the delivery of social services. The impacts of Outer Continental Shelf–related activities were not uniform throughout the coastal Gulf of Mexico. The communities hit hardest by the Outer Continental Shelf boom were those traditional settlements, which had become staging areas for offshore activities (Morgan City, Berwick, Patterson, Houma, Larose, Golden Meadow, and Grand Isle), and managerial centers for offshore activities (primarily Lafayette). Some of the traditional port cities of Louisiana and Texas (New Orleans, Baton Rouge, Lake Charles, Port Arthur, Houston [see especially Feagin 1985], and Galveston) also experienced growth of both staging and managerial activities, but because of their larger size and more diversified economies, the impact was less than in the rural coastal communities on the deltaic plain.

Particularly hard hit were some of the communities in the eastern part of St. Mary Parish (Morgan City, Berwick, Patterson) and the eastern part of Terrebonne Parish (Gramling and Brabant 1984, 1986a; Gramling and Freudenburg 1990). Housing became critically short and real estate prices escalated, as demand far exceeded supply. The provision of public utilities and services, water supplies, sewage treatment, utilities (Johnson 1977; Durio and Dupuis 1980), transportation (Stallings and Reilly 1980), recreational facilities (Reilly 1980), and medical facilities (Gramling and Joubert 1977) lagged behind the population growth throughout the 1960s and 1970s. One of the hardest hit community services was public education. The educational system in Louisiana has never been sufficiently funded. The influx of students into systems that were already marginal at best led to teacher shortages, overcrowding, and scarce supplies. Because demand increased more rapidly than the tax base, there was little way for the supply of these services to catch up (Gramling and Reilly 1980; Brabant 1984).

The offshore activity attracted transient labor for work offshore, and the economic growth in general attracted significant numbers of the transient, chronically unemployed (Brabant 1993a, 1993b; Brabant and Gramling 1985, 1991). Labor camps, which provided lodging, food, and job contacts for exorbitant prices, were in operation in the Morgan City area. Shelters and social service agencies, which provided basic human services (food, clothing, shelter), were often taxed beyond their capacity (Brabant 1993b). Crime rates rose. By the mid-1970s the violent crime rate in Morgan City ranged from two to three times the average rate for cities of comparable size in the United States. By the mid-1970s portions of coastal Louisiana exhibited many of the characteristic stresses and strains associated with the classic boom town syndrome (Gramling and Brabant 1986a; Gramling and Freudenburg 1990; Seydlitz et al. 1993a). The region received national attention during this period, and articles appearing in major tabloids exacerbated the in-migration of transient labor.

The Embargo

The 1973–1974 oil embargo by the Arab members of OPEC against the United States and the Netherlands in retaliation for their support for Israel during the 1973 round of the Arab-Israeli war became a major stimulus for offshore development in the Gulf and in the remainder of the United States. While crude oil production on the Gulf Outer Continental Shelf had actually peaked and begun to decline by this time (although natural gas production was still climbing; Manuel 1984), offshore activity continued to grow. By the time of the embargo the offshore industry was a worldwide enterprise, although the Gulf of Mexico continued to be a dominating influence for the next decade and a half. Two factors, noted in chapter 4, directly affected Outer Continental Shelf operations: (1) the dramatic rise in the price of crude oil following the embargo provided increased economic incentive for exploration and development on the OCS and (2) Project Independence, President Nixon's proposed quick fix for the shortages and price rises. Outer Continental Shelf sales resumed in the Pacific, the Atlantic, and in Alaska following the initiation of Project Independence. Table 3.2 (chapter 3) shows the marked increase in exploratory and production drilling (the most labor-and resource-intensive phases) following the 1973–1974 embargo. As a result of the new impetus provided by the skyrocketing price of crude oil and an aggressive federal leasing system, the offshore industry expanded its operations. In the United States, in spite of federal encouragement, this expansion was limited primarily to the Gulf of Mexico.

Table 5.2 shows the pattern of lease sales in the Gulf of Mexico over the history of leasing on the Outer Continental Shelf. While the majority of the early sales were off the Louisiana deltaic plain (see Fig. 5.1), by the mid-1960s sales had increased in western Louisiana and the shallow areas off Texas. Following the embargo and the in-

Table 5.1. Population and Growth Rates, Coastal Louisiana and Texas

PARISH/COUNTY	POPULATION OF SELECT COASTAL ZONE PARISHES 1940–1990						PERCENTAGE CHANGE 1940–1990				
	1940	1950	1960	1970	1980	1990	40/50	50/60	60/70	70/80	80/90
Louisiana											
Assumption	18,541	17,278	17,991	19,564	22,084	22,753	-6.8	4.1	8.7	12.9	3.0
Calcasieu	56,506	89,635	145,475	145,415	167,223	168,134	58.6	62.3	0.0	15.0	0.5
Cameron	7,203	6,244	6,909	8,194	9,336	9,260	-13.3	10.7	18.6	13.9	-0.8
Iberia	37,183	40,059	51,672	57,397	63,752	68,297	7.7	29.0	11.1	11.1	7.1
Jefferson	50,427	103,873	208,769	337,568	454,592	448,306	106.0	101.0	61.7	34.7	-1.4
Lafayette	43,941	57,743	84,656	111,745	150,017	164,762	31.4	46.6	32.0	34.2	9.8
Lafourche	38,615	42,209	55,381	68,941	82,483	85,860	9.3	31.2	24.5	19.6	4.1
Orleans	494,537	570,445	627,525	593,471	557,515	496,938	15.3	10.0	-5.4	-6.1	-10.9
Plaquemines	12,318	14,239	22,545	25,225	26,049	25,575	15.6	58.3	11.9	3.3	-1.8
St. Mary	31,458	35,848	48,833	60,752	64,253	58,086	14.0	36.2	24.4	5.8	-9.6
Terribonne	35,880	43,328	60,771	76,049	93,393	96,982	20.8	40.3	25.1	22.8	3.8
Vermillion	37,750	36,929	38,855	43,071	48,458	50,055	-2.2	5.2	10.9	12.5	3.3
Louisiana	2,363,880	2,683,516	3,257,022	3,643,180	4,205,900	4,219,973	13.5	21.4	11.9	15.4	0.3

Table 5.1. continued

PARISH/COUNTY	POPULATION OF SELECT COASTAL ZONE PARISHES 1940–1990						PERCENTAGE CHANGE 1940–1990				
	1940	1950	1960	1970	1980	1990	40/50	50/60	60/70	70/80	80/90
Texas											
Brazoria	27,069	46,549	76,204	108,312	169,587	121,862	72.0	63.7	42.1	56.6	-28.1
Calhoun	5,911	9,222	16,592	17,831	19,574	19,053	56.0	79.9	7.5	9.8	-2.7
Cameron	83,202	125,170	151,098	140,368	209,727	260,120	50.4	20.7	-7.1	49.4	24.0
Chambers	7,511	7,871	10,379	12,187	18,538	20,088	4.8	31.9	17.4	52.1	8.4
Galveston	81,173	113,066	140,364	169,812	195,940	217,399	39.3	24.1	21.0	15.4	11.0
Harris	528,961	806,701	1,243,158	1,741,912	2,409,547	2,818,199	52.5	54.1	40.1	38.3	17.0
Jefferson	145,329	195,083	245,659	244,773	250,938	239,397	34.2	25.9	-0.4	2.5	-4.6
Matagorda	20,066	21,559	25,744	27,913	37,828	36,928	7.4	19.4	8.4	35.5	-2.4
Nueces	92,661	165,471	221,573	237,544	268,215	291,145	78.6	33.9	7.2	12.9	8.5
Orange	17,382	40,567	60,357	71,170	83,838	80,509	133.4	48.8	17.9	17.8	-4.0
San Patricio	28,871	35,842	45,021	47,288	58,013	58,749	24.1	25.6	5.0	22.7	1.3
Texas	6,414,824	7,711,194	9,579,677	11,198,655	14,229,191	16,986,510	20.2	24.2	16.9	27.1	19.4
U.S.	131,669,275	150,697,361	179,323,175	203,225,299	226,549,448	248,709,873	14.5	19.0	13.3	11.5	9.8

Source: U.S. Department of Commerce, Bureau of the Census 1940, 1950, 1960, 1970, 1980, 1990.

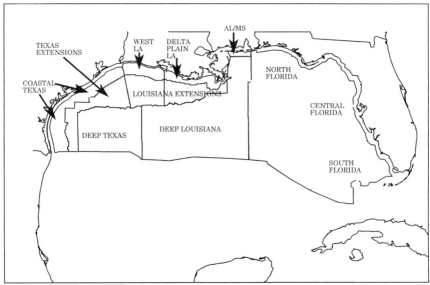

Fig. 5.1. Gulf Offshore Areas.

creased federal offerings, movement into these areas increased, along with sales in deeper water in both Louisiana and Texas.

Looking at the second panel in Table 5.2 we can see, however, although sales moved into other regions in the Gulf of Mexico, that the primary area of actual development (i.e., siting of production platforms) continued to be off the deltaic plain with accompanying movement into western Louisiana waters. This geographic trend in lease sales and development was paralleled by construction of offshore production platforms.

Table 5.3 shows the number of offshore production platforms under construction worldwide in 1975, 1978, and 1981. Several important trends are noticeable. First, like sales and development, by 1975 the offshore fabrication industry had begun to move out of the Louisiana deltaic plain into Texas, a trend that accelerated over the next decade and a half. Second, throughout this period following the embargo until the peak of the offshore boom in 1981, the Gulf of Mexico continued to be the site of almost half of the world offshore production platform construction. Third, throughout this period Louisiana dominated the market for fabrication of production platforms within the Gulf.

In a similar fashion, Louisiana dominated the fabrication of off-shore support vessels, becoming a world leader in the construction of literally thousands of crew boats, supply boats, push boats, tow boats, and various other more specialized offshore vessels (*Ocean Industry* 1974; Tubb 1978).

In spite of this dominance in offshore construction and support, the geography of the deltaic plain did provide one limitation. Between the Calcasieu River and the mouth of the Mississippi, a nautical distance of approximately 200 miles, and containing the area of heaviest offshore development, historically there have been no deep-water ports. The only exceptions are the Atchafalaya River, which has a twenty-foot-deep-channel as far inland as Morgan City and more recently Bayou LaFourche. With the Atchafalaya River, the actual frontage on the river is extremely limited with most of the connections to the major fabrication yards, both in Morgan City and throughout the deltaic plain, being via the Gulf Intracoastal Waterway, which is only guaranteed to twelve-foot depths. Bayou LaFourche has a twenty-foot depth for only several miles inland from its mouth, and this only for the last decade. This does not pose a problem for the fabrication of steel-jacketed production platforms, which are fabricated on their side, loaded onto a barge(s) of less than twelve-foot draft, and taken offshore. It does mean that no shipbuilding facilities have emerged.

Unlike production platforms, exploratory drilling rigs must be fabricated and floated to their initial destination, and virtually all of the deep-water rigs require more than twelve feet of water. As the offshore rig technology evolved, by necessity, construction moved out of Louisiana. By 1976, much of the U.S. construction of exploratory rigs was taking place in Texas and Mississippi (primarily Pascagoula and Vicksburg), a general trend that continued into the 1980s (*Ocean Industry* 1976; Tubb 1977). In addition to direct activities occurring in the Gulf, movement into many of the "frontier" regions, under the new push for offshore development on federal lands, continued to be supported from the Gulf. Experienced offshore workers and even decks for production platforms off California continued to come from the Gulf (Gould et al. 1991).

In spite of the incredible rate of growth of the economy and the freak entrepreneurial conditions (noted in chapter 4) under which the growth occurred (Gramling and Brabant 1986a; Gramling and

Table 5.2. Gulf Area by Leases Sold

	DELTA LA[a]	DELTA EX-LA[b]	EAST LA[c]	WEST LA[d]	DEEP LA[e]	COAST TX[f]	EX TX[g]	DEEP TX[h]	AL MS[i]	NO FL[j]	MIDL FL[k]	SO FL[l]
						GULF AREA BY TOTAL LEASES SOLD						
Before 1954	169	0	13	96	0	97	0	0	0	0	0	0
1954–1959	115	0	0	98	0	46	0	0	0	0	0	0
1960–1964	219	108	37	173	0	59	0	0	0	0	0	0
1965–1969	110	45	56	31	0	124	36	0	0	0	0	0
1970–1974	112	133	42	272	52	23	169	14	3	39	26	0
1975–1979	117	59	28	229	24	157	121	3	0	9	16	14
1980–1984	267	218	137	425	576	499	206	146	47	92	7	89
1985–1989	267	259	108	473	1,293	465	261	207	33	78	8	13
1990–1993	179	154	79	336	456	252	99	57	11	74	19	0
Total	1,555	976	500	2,133	2,401	1,722	892	427	94	292	76	116

Table 5.2. continued

GULF AREA BY LEASE DATE WITH AT LEAST ONE PLATFORM INSTALLED BY 1992

	DELTA LA[a]	DELTA EX-LA[b]	EAST LA[c]	WEST LA[d]	DEEP LA[e]	COAST TX[f]	EX TX[g]	DEEP TX[h]	AL MS[i]	NO FL[j]	MIDL FL[k]	SO FL[l]
Before 1954	80	0	5	31	0	2	0	0	0	0	0	0
1954–1959	41	0	0	15	0	3	0	0	0	0	0	0
1960–1964	100	33	11	42	7	0	0	0	0	0	0	0
1965–1969	36	8	15	8	0	4	2	0	0	0	0	0
1970–1974	49	47	9	115	7	6	59	3	0	0	0	0
1975–1979	43	20	14	83	4	48	27	0	0	0	0	0
1980–1984	86	26	50	117	13	110	17	1	17	0	0	0
1985–1989	27	14	12	49	6	55	8	0	5	0	0	0
1990–1993	2	1	2	1	1	1	0	0	0	0	0	0
Total	464	149	118	461	31	236	113	4	22	0	0	0

Source: Minerals Management Service 1993. (a) Deltaic plain Louisiana (original offshore areas); (b) Extension areas off the deltaic plain (e.g., Grand Isle South edition); (c) Louisiana, mouth of the Mississippi and eastward; (d) Louisiana west of Vermillion Bay; (e) Deep water Louisiana (south of the original "south editions" to the limits of U.S. waters); (f) Coastal Texas (original offshore areas); (g) Extension areas off Texas; (h) Deep water off Texas; (i) Off Alabama and Mississippi; (j) Northern and Western Florida offshore areas (including Destin Dome and DeSota Canyon areas); (k) Middle Florida (including the Florida Middle Grounds); (l) South Florida (including Pulley Ridge).

Table 5.3. Fixed Platforms Under Construction—January 1975, March 1978, and March 1981

LOCATION	TOTALS			PERCENTAGE		
	1975	1978	1981	1975	1978	1981
California	1	0	2	0.6	0	0.6
Louisiana	61	55	94	37.4	37.4	30.4
Texas	9	20	52	5.5	13.6	16.8
Outside U.S.	92	72	161	56.4	49.0	52.1
Total worldwide	163	147	309	100	100	100

Source: Davis and Place 1983.

Freudenburg 1990), the direction of the growth was one that would seemingly be an almost textbook example of regional economic development success. As Lovejoy and Krannich (1982) have noted in summarizing the economic development debate, often the "success" of a particular economic development scenario hinges at least as much on the ability of the region to capture spin-off and spill-over activities, as on the primary development. These linked activities, both upstream (those that supply the primary activity) and downstream (those that use the product produced by the primary activity), can generate jobs and capital, which supplement the primary activity and result in a more diversified form of development.

Since the Gulf of Mexico was the primary area for offshore development throughout the 1950s, 1960s, and much of the 1970s, and since the problems faced were local, most of the technological and human capital development was, by necessity, a local phenomenon. Because much of the support for these activities had to be local, as a consequence, the capture of spin-off was almost inevitable. The extent to which this happened can be seen by examining offshore-related employment (often a more important indicator of development than population growth, Gramling 1992) in one of the areas most closely tied to offshore development, the Louisiana deltaic plain, west of the Mississippi River delta. Table 5.4 shows total employment, employment in oil and gas extraction, and employment in three of the sectors most directly related to offshore activity: metal fabrication (platform construction), ship and boat building and repair, and water transportation for Lafayette, Lafourche, and St. Mary and Terrebonne parishes. These are certainly not the entire employment associated with offshore oil, and, in fact, examination of only these limited sectors misses much of the employment directly related to offshore development (e.g., offshore catering, marine diesel mechanics, hot shot drivers, etc.) and most of the secondary and tertiary employment (e.g., land transportation, construction, insurance, accounting, etc.). Nevertheless, the marked increase in employment in only these sectors in the late 1950s and early 1960s, as the offshore system of development was emerging, and the steady proportion of total employment (up to 30%) through the mid-1980s are indicative of the extent to which the social and economic environments along the coastal Gulf were affected by offshore trends. The growth of employment in oil and gas extraction exploded following the settlement of the litigation between Louisiana and the federal government and the resumption of an aggressive federal leasing program in 1959. While employment grew slowly during the 1964–1974 decade, it again jumped significantly following the 1973–1974 embargo. Employment in the linked sectors grew more steadily throughout the 1959–1979 period and by 1974 had actually exceeded employment in the primary sector in two of the parishes.

REBUILDING THE ENVIRONMENT

The growth was vigorous in both the primary and the linked industrial sectors through the 1970s. However, that growth was predicated on an extractive economic activity. As noted in chapter 1, pithole-like volatility is characteristic of extractive activity both because of eventual resource exhaustion and because of the way in which extractive enterprises affect the local environment (Bunker 1984, 1989; Gramling and Freudenburg 1990; Freudenburg and Gramling 1992a). Because extractive enterprises must locate in proximity to the resource, they cannot necessarily locate near existing development and take advantage of shared labor supplies and support sectors. As a result, they must often rebuild the local environment (social, economic, and physical) to provide support for the extractive activity. This rebuilding of the environ-

Table 5.4. Employment in Oil and Linked Sectors Coastal Louisiana

YEAR	TOTAL	OIL/GAS EXTRACTION	PERCENTAGE OIL/GAS EXTRACTION	LINKED SECTORS[a]	PERCENTAGE LINKED SECTORS
Lafayette Parish					
1959	14,426	621	4.30	0	0.00
1964	18,487	2,747	14.86	265	1.43
1969	25,071	3,627	14.47	188	0.75
1974	35,441	2,362	6.66	859	2.42
1979	62,572	7,620	12.18	1,303	2.08
1984	76,644	10,073	13.14	1,430	1.87
1989	65,615	5,710	8.70	781	1.19
Lafourche Parish					
1959	7,497	608	8.11	655	8.74
1964	10,269	2,149	20.93	1,106	10.77
1969	10,610	1,122	10.57	1,763	16.62
1974	13,622	940	6.90	3,149	23.12
1979	18,605	1,081	5.81	5,638	30.30
1984	17,418	845	4.85	4,135	23.74
1989	15,137	175	1.16	3,588	23.70
St. Mary Parish					
1959	7,750	755	9.74	922	11.90
1964	12,139	2,943	24.24	1,762	14.52
1969	16,477	3,219	19.54	2,537	15.40
1974	19,374	1,812	9.35	4,420	22.81
1979	31,212	3,552	11.38	8,436	27.03
1984	25,181	1,373	5.45	8,095	32.15
1989	17,519	981	5.60	4,146	23.67
Terrebonne Parish					
1959	8,032	165	2.05	507	6.31
1964	12,528	2,821	22.52	945	7.54
1969	16,443	3,926	23.88	1,310	7.97
1974	26,331	6,641	25.22	3,573	13.57
1979	33,598	6,294	18.73	3,976	11.83
1984	31,663	4,904	15.49	2,495	7.88
1989	28,143	3,834	13.62	1,651	5.87

Source: U.S. Department of Commerce 1959, 1964, 1969, 1974, 1979, 1984, 1989. (a) water transportation, metal fabrication, ship and boat building and repair.

ment has several consequences (Freudenburg and Gramling 1992a). First, the specialized development (social, economic, and physical) surrounding the extractive activity is often not transferable to new activities, and thus flexibility is lost, as local human and financial capital focus ever more narrowly on the primary extractive and support sectors. Second, the creation of the new support sectors may use up or destroy local resources. Third, the existence of high-paying jobs in the extractive sector makes the competition for labor keen and thus the introduction of alternative economic activities difficult. When the new extractive activity ceases or declines, the local area is more specialized than before the extractive activity started, and there is less of the local resource base available. Thus, the development cycles associated with extractive activities can lead to "overadaptation" and inflexibility (Gramling and Freudenburg 1990, 1992a; Freudenburg and Gramling 1992a).

When the OCSLA passed in 1953, Lafayette was the distribution center of the traditional "Acadiana" region, having recently been transformed from a railroad town to the center of a highway network (Gramling 1983). Morgan City was the self-proclaimed "shrimp capital of the world," and most of the remainder of the region was primarily oriented toward agriculture or the harvesting of renewable resources: shrimp, fish, crawfish, and so on. (Comeaux 1972). Over the next three decades, offshore activities gradually came to constitute the most important primary sector of the economy (both in the Acadiana area and eastward along the coast), and secondary and tertiary support sectors developed in response to the growth opportunity.

Brabant sums up the change dramatically:

> Alexander Godunov, Mikhail Baryshnikov, Rudolph Nuriyev—names associated with stages in New York, San Francisco, possibly Chicago, Washington D.C., maybe New Orleans—but Lafayette, Louisiana? Yet between 1982 and 1986, each of these internationally acclaimed ballet superstars appeared on the stage of the Lafayette Municipal Auditorium. . . . How could this be? The answer is simple—oil. (Brabant 1993a: 161)

The change was pervasive. Trade schools altered their offerings (in response to demand) to teach ever more esoteric skills, marketable primarily offshore and in offshore support sectors (cf. Gramling and Reilly 1980). High school education (or less) was sufficient for good wages with these types of skills, and, as a result, many entered the labor force early but with few transferable skills. The director of an area economic development association noted this clearly in retrospect:

> The downside to the oil industry as I see it from economic development has been the deterrent in education. Because so many of our people—high school students, drop-outs, those types, were able to get such good jobs—as a roustabout working in the oil industry—that they didn't finish their education. They've come back. They've either been sitting on the unemployment roll or we've convinced them to go into other types of training—skilled labor training. (Freudenburg and Gramling 1994a: 40)

Even with no skills, labor in offshore development or in many of the support sectors was well paid. The opportunities were extensive, since the primary activities, drilling and production, were massive economic undertakings, and, as noted, cost was a secondary consideration to time.

Thus, new investment (some of it massive) centered around the needs of the offshore sector. Fabrication yards sprang up on the banks of local bayous as offshore production platforms, offshore drilling rigs, support vessels, and metal fabrication of all types were in great demand. In order to attract the fabrication and construction industries associated with offshore activities and thus produce local jobs, communities approved long-term bond issues for the construction of the marine equivalent of local industrial parks (small ports) contiguous to the waterfront near the community, committing local resources to the continuation of the offshore sector. In addition to fabrication yards, offshore catering, drilling mud service, oil field tubing (drill pipe, casing), transportation (water, land, and air), and various types of specialized equipment (compressors, tanks, etc.) and instrumentation (down hole logging) companies were chartered and thrived during the offshore boom. New development was not the whole picture, as conversion of existing facilities, such as dock space, also significantly altered the environment. By the mid-1970s, Morgan City, "shrimp capital of the world," had no resident shrimp fleet and no operating shrimp processing plants.

Under these circumstances, local social and economic systems adapted and did so at a variety of levels. On an individual level, people made career decisions based on the expectation that past trends would continue. Seventeen year olds dropped out of high school to gain and use (often esoteric) skills in the support and fabrication sectors surrounding offshore activities. Those who graduated often pursued specialized types of skills (associated with the offshore and support sectors) at the expense of other skills or more flexible higher education (Seydlitz et al. 1993b).

The incentive to move into the offshore support sectors was high. Employment offshore and in the oil-related sector in general paid well. Table 5.5 compares income distribution for those who work in a concentrated scheduling format (e.g., seven and seven) across sectors of the economy directly related to oil production and sectors not directly related to oil production. In general it is quite clear that both working shifts and working in sectors of the economy directly related to the production of oil influenced income positively.

In addition to changes in individual careers, adaptation occurred within organizations. At the small business level, businesses tended toward specialization. Mechanic shops became marine diesel repair facilities (with considerable investment in tools and equipment to make the transition). New specialty businesses opened to take advantage of the growing opportunities (offshore catering services, hot shot drivers[3], etc.). Both of these trends happened in an economic environment of such unparalleled prosperity that good business practices were not necessary for success. With profit margins high enough, inventory control, billing, machine and equipment layout, and so on could be marginal, and still enterprises prospered (Freudenburg and Gramling 1992a). As one informant put it "people made money in spite of themselves, and unfortunately came to believe that they were good businessmen."

Adaptation also occurred at the regional level, as the interaction between the resources associated with human and social capital,

Table 5.5. Annual Income by Industrial Sector and Shift Work, 1980

| | SHIFT WORK | | | |
| | NO | | YES | |
ANNUAL INCOME	NON-OIL	OILFIELD	NON-OIL	OILFIELD
0-4,999	147	8	8	6
5,000-9,999	293	30	5	32
10,000-14,999	168	39		59
15,000-19,999	80	54	2	120
20,000-24,999	78	20	2	100
25,000-29,999	31	11	1	42
30,000-34,999	12	8		19
Over 35,000	31	17		12
Totals	840	187	18	390

Source: Gramling 1980a.

skills, knowledge, experience, teamwork, and networks of supply and distribution and the physical capital of buildings, equipment, and other physical infrastructure developed quickly and in response to the potential for profit. The 3,600 advertised business activities directly connected to oil and gas activities that Davis and Place (1983) located in coastal Louisiana were indicative of this trend.

Simply put, the northern Gulf of Mexico was by 1980, hands down, the most developed, and impacted, area in the world with regard to offshore oil and gas activities. It was also the most specialized area in the world with regard to offshore activities, raising the specter, as Freudenburg and Gramling (1992b) noted, that with extractive economies, even those with the extensive development of linkages, the region could find all of its linked economic eggs in one falling basket. As U.S. demand for petroleum products began to fall in the late 1970s, resulting in a crash in oil prices in the mid-1980s, this lesson became quite apparent.

In addition to individuals, businesses, and the infrastructure associated with communities, overadaptation may also be characteristic of smaller social systems, such as families. As noted in chapter 3, Forsyth and Gramling (1987) examined the adaptation strategies that families utilize to adapt to the periodic absence of one member of the family (usually the male) in situations, such as those associated with employment in the offshore oil sectors (Gramling and Forsyth 1987; Forsyth and Gauthier 1991, 1993; Storey et al. 1986; Fuchs et al. 1981). Their basic conclusion was that the resulting interaction within families, while different from families experiencing more traditional forms of work scheduling, was not evidence of pathology but of adaptation. Adaptation to these altered forms of work scheduling is problematic and is one of the most frequently mentioned problems with work in the "oil patch." The point is that certain forms of development (here extensive offshore oil and gas development) can lead to altered forms of interaction within the family (such as minimal involvement of the male in day-to-day activities), which are maladaptive if the situation changes and offshore work is no longer available.

BEGINNING OF THE END

Following the oil embargo the price of crude oil rose steadily throughout the 1970s, and the rising value of the commodity had two predictable results. First, motivated by an obvious profit motive, the major players in the world oil market sought to increase their production and were quite successful. Figure 5.2 shows the increase in world production during this period. In 1979 world oil production peaked at levels that have not occurred since.

The second predictable result of skyrocketing oil prices was a fall in consumption. Oil is what economists call an elastic commodity. Unlike commodities that are driven by a fixed need, like insulin, and show little variation in consumption as price changes, oil consumption (or more properly consumption of its refined products) varies considerably with the price of the product. However, in the case of petroleum products, this decline in consumption shows up gradually as consumers replace their automobiles with more fuel-efficient models, buy better insulated homes (or insulate the ones they own), and make a host of other travel- and energy-related decisions. The average mileage for automobiles per gallon of gasoline was 13.1 in 1973, by 1985 it had risen to 17.9, a 36.6 percent increase (Beck 1988) As a result of this trend, by 1981

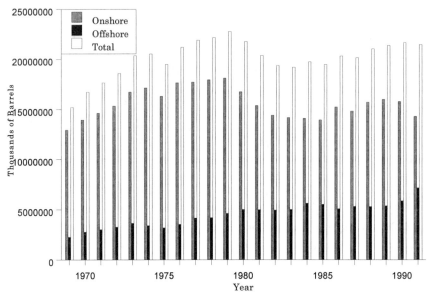

Fig. 5.2. World Oil Production, 1969–1991. *Source*: API 1993.

consumption of petroleum products in the United States had fallen to below pre-embargo levels.

With the growing production of oil and the falling consumption of the refined products, the stage was set for the collision of these two trends, and that collision came in early 1986. While the price of oil fluctuated between 1982 and 1985, by December of 1985 oil was still $24.51 per barrel, but by July of 1986, it was $9.39. The growth-related problems in the coastal Gulf of Mexico began to moderate somewhat when crude oil prices began to fall in 1982. However, with the crash in crude oil prices in 1986, it became evident that the integration of the offshore industry into the local communities had resulted in a significant modification of the social, economic, and physical environment and a whole new set of problems.

The decline in prices on the world oil market demonstrated the vulnerability of the offshore oil- and gas-related activities, and accompanying development, to trends in the commodities market and the extent to which coastal Louisiana and Texas had become dependent on these activities (see Manuel 1980; Gramling and Freudenburg 1990). Over 90 percent of the variance in total employment in some coastal Louisiana parishes over the last two decades can be statistically explained using only national and international indicators of the world oil market, world rig count, average annual price of oil, and so on (Gramling and Freudenburg 1990). This means that in these coastal parishes, events completely beyond their control dominate their economies to an extent rarely seen in development scenarios. What has also become clearly evident is that much of this development and many of these skills are not directly transferable.

THE BUST

The fall in crude oil prices in late 1985 and early 1986 hit coastal Louisiana and Texas hard. Much of the investment and accumulation in both financial and human capital, associated with the rebuilding of the physical, social, and economic environment, was lost. The reaction along the coastal Gulf of Mexico was near panic. Unemployment in the most heavily affected coastal parishes in Louisiana, which had remained in the 4 to 5 percent levels for decades, exceeded 20 percent in some cases by the end of 1986. Table 5.4 also shows the downturn in employment in the most heavily affected parishes in Louisiana.

The region received national attention again due to the local recession, the effects of which were sufficient to bring Senate hearings to the region (Senate Committee on Energy and Natural Resources 1987). A more poignant local indicator was when the formerly exclusive Petroleum Club in Lafayette was forced to open its membership to nonoil industry members (*The Times of Acadiana* 1993). Once again social services came under pressure but for different reasons. Whereas before local organizations were hard pressed to respond to the needs of migrants looking for jobs, now these same local organizations were finding that their clients were formerly well-employed local residents, newly unemployed and looking for jobs.

Brabant quotes one informant:

We had massive unemployment. We had people standing in line all the way out the door and into the streets. We didn't have near enough staff to service the people coming in. We were hiring temporary people. . . . We had engineers, people with Ph.D.'s, geophysicists, geologists, people making upwards to $250,000 a year coming in and filing claims for unemployment insurance for $205 a week. . . . If you or a member of the family are a stockholder in a company, you must prove that the company has been dissolved. People had board meetings at the interviewer's desk to dissolve their companies right here in the office so they could qualify for the unemployment insurance. It was really a hard thing to look out and see. (Brabant 1993a: 184–185)

The bust forced many to leave to look for work (Table 3.3), and since the same people who migrated in are not necessarily the ones to leave, this often resulted in the separation of extended families whose descendants had been in the area for over two centuries.

This flight of financial and human capital marks the beginning of the inevitable disassembly of the offshore system and its onshore support network in the Gulf of Mexico. Ironically, because of the out-migration, in some cases new development initiatives in the last several years have found skilled labor in short supply.

Chapter 6

Rising Political Controversy

BACKGROUND TO CONFLICT

During the same time period that development of the Outer Continental Shelf in the central and western Gulf of Mexico was generating an incredible economic boom, elsewhere the proposed development generated organized resistance to the Department of Interior's program. Outside of the central and western Gulf (as noted in chapter 4), Outer Continental Shelf leasing was not a popular activity. The initial sales in Alaska and the resumption of sales in California and off the East Coast in the mid-1970s met with reactions from various interest groups and state governments and counterreactions from the supporters of offshore oil. During the more visible period of the conflict, Governor Edwards of Louisiana threatened to shut off the flow of oil to consumption regions that continued to resist exploration (offshore) and refining in their states (Engler 1977).

On the East Coast, concerns over the renewed leasing program led to the formation of the Mid-Atlantic Governors Coastal Resource Council (Wilson 1982). Given the possibilities of vastly increased Outer Continental Shelf leasing, the Mid-Atlantic Governors Coastal Resource Council began to investigate the possibilities. Representatives sent to examine the effects of Outer

Continental Shelf development in the Gulf of Mexico were apparently appalled by what they saw. In 1975 the governors passed a resolution, which noted their concern:

> WHEREAS, it is clear from the experience of other states adjacent to offshore operations and from the experience of countries bordering North Sea operations, that such activities can have a severe and detrimental impact on the economy, environment, and social structures of affected areas (cited in Wilson 1982: 77)

They called on Congress and the president to undertake "with the full and active involvement of the States, . . . a detailed and accurate analysis of the onshore impacts of the program" (Wilson 1982: 77). The following year Alaska went to court to try to stop a lease sale in the Gulf of Alaska (Wunnicke 1982), and Congress began to get seriously involved.

Soon after the announcement of Project Independence, attempts to amend the Outer Continental Shelf Lands Act were introduced in Congress. The states' resistance to the Outer Continental Shelf leasing program increased the pressure to act, and in 1978, after four years of debate, Congress passed the Outer Continental Shelf Lands Act Amendments (OCSLAA). Congress was concerned that the Outer Continental Shelf leasing and regulation was essentially a closed decision process involving the secretary of interior and industry. Thus, one of the purposes of the amendments was to open the decision-making process to a wider audience and thereby increase public confidence in this federal activity (Legislative History, P.L. 95-372, p. 54).

In general, the amendments intended to expedite the development of the resources of the Outer Continental Shelf in a fashion so as "to balance orderly energy resource development with protection of the human, marine, and coastal environments." In addition, the amendments intended input from, and protection for, state and local communities, and further required that sufficient information be collected to assess and monitor the impact of Outer Continental Shelf activities (see Appendix for relevant text, see Krueger and Singer 1979; Jones et al. 1979 for detailed analysis of the amendments).

While the Department of the Interior took the first purpose of the act, "expedited exploration and development of the Outer Continental Shelf in order to achieve national economic and energy policy goals, assure national security, reduce dependence on foreign sources, and maintain a favorable balance of payments in world trade" (43 U.S.C. § 1802) quite seriously, a number of states felt that the input for state and local governments and the information collection for assessment and monitoring were lacking. Primary among these were California and Florida.

In California, the line in the sand for state-federal conflict over the Outer Continental Shelf came with proposed sale 53 in the Pacific. Unlike previous sales, which were concentrated in one geographic region, proposed sale 53 called for nominations of tracts from the Santa Barbara Channel area north to the Oregon state line. Protests against the sale started with the call for nominations in November of 1977 and grew with the announcement of proposed tracts a year later. The Coalition on Lease Sale 53, representing a score of interest groups, was formed to oppose the sale, and throughout 1978 and 1979 Interior faced the stiffest opposition of any previous sale. At this same time, opposition was also mounting against the first five-year leasing schedule, which was required under the Outer Continental Shelf Lands Act Amendments and which was announced by Interior in 1979. California and Alaska challenged the plan on grounds that it violated section 18 of the OCSLAA and eventually won in Washington D.C. District Court (Fitzgerald 1987). When the protests extended into 1980, then Secretary of Interior Andrus withdrew the entire northern and central California portion of the sale and canceled a number of controversial tracts near the Santa Maria Basin in the south (Kaplan 1982). However, the issue passed from Andrus's hands with the election of Ronald Reagan in 1980. Reagan's approach to energy policy was to let the market determine energy use (Freudenburg and Gramling 1994b; Riposa 1989), and opening the market became a major consideration.

Shortly after the Reagan administration was sworn in, the new Secretary of Interior James Watt, in his first public policy statement, announced he was reversing Andrus's decision and restoring sale 53 to its original offerings. California, joined by nine other coastal states and a number of cities and counties, again went to court. This time the suit contended that the sale did not meet consistency under the CZMA, relighting the controversy surrounding sale 48. When the state won the initial round in court, Watt withdrew the contested offerings.

In another move to expand Outer Continental Shelf leasing in 1981, Watt proposed an amended five-year leasing plan with a new strategy that opened the leasing process to entire planning areas (e.g., the western Gulf of Mexico), rather than the much more limited selection of tracts, which had been in effect since the first federal lease sale in 1954 (Gould et al. 1991). The proposal for "area-wide" leasing met with almost unanimous opposition from the states,[1] increased opposition to the Outer Continental Shelf program, and resulted in renewed political pressure at both the state and federal levels. Again California, joined by Alaska, Washington, Oregon, and Florida, went to court. While this whole process was under litigation, Watt combined all of the leasing, regulation, and study production functions pertaining to the Outer Continental Shelf program under one agency within the Department of Interior, thereby creating Minerals Management Service and concentrating the growing pressure against the Outer Continental Shelf leasing program on one agency.

The states were not successful in their litigation, and the following year Watt initiated "area-wide" leasing which drastically increased the acreage offered for lease. In the Gulf, prior to area-wide leasing, the record number of acres offered (sale 37, February 1975) was 2,870,344. The first area-wide sale offered 37,867,762 acres, and the first area-wide sale in the eastern Gulf in January of 1984 resulted in leases being sold in the area south of the 26 parallel off Florida (near the Florida Keys), over the protests of Governor Bob Graham. This later escalated into a full-scale political battle between Graham's successor Bob Martinez and Watt's successor Hodel (see chapter 7).

The primary stated motivation behind Nixon's initial expansion of the federal Outer Continental Shelf leasing program, both in terms of movement into new "frontier" areas and increased production in the Gulf, was energy independence, and during Watt's tenure at Interior, this perspective became *the* focus of the program, though given the growing budget deficit, there were those who doubted his sincerity.

> Current U.S. energy demands are met primarily by domestic and foreign fossil fuel. Since the 1973 Arab oil embargo it has become increasingly apparent that our nation must become less dependent on foreign imports, lessen our vulnerability to supply economics and supply interruptions, and prepare for the time when oil production approaches its capacity limitation. In 1978, Congress and the President mandated the DOI to engage in "expedited exploration and development of" the OCS in order to "assure national security, reduce dependence on foreign sources, and maintain a favorable balance of payments in world trade."[2] The Secretary [Watt] has stated that "we honor that mandate, and until there is other direction, it will be our foremost guideline in all OCS activity." (Minerals Management Service 1984: 3)

CONGRESSIONAL INTERVENTION

In 1983, the Supreme Court, ruling on the issue raised by California, overturned lower courts and ruled that the lease sale itself did not cause impacts and therefore did not come under the consistency clause in the CZMA. This decision, coupled with the one allowing area-wide leasing to proceed, seemingly left Watt in a strong legal position to move ahead with a more aggressive Outer Continental Shelf development program, but legal positioning was becoming irrelevant, as the coastal states moved to bring pressure in Congress.

In 1982, Congress inserted prohibitions into the Department of Interior's appropriation forbidding the expenditure of funds for leasing activities for portions of the East and West coasts' Outer Continental Shelf (Farrow 1990). In effect, this back-door process circumvented the decision-making process within Interior and increasingly shut down the Outer Continental Shelf outside of the western and central Gulf of Mexico to Outer Continental Shelf leasing. Although Congress did not initially delete the area in

Florida south of the 26th parallel from lease sales, later Congressional appropriations language imposed a moratorium on exploratory drilling until two stipulations were met: (1) no exploratory drilling activities would be approved by the Department of Interior until three years of physical oceanographic and biological research data had been collected and, (2) leasees would be required to perform biological surveys prior to initiation of drilling operations and work with the Department of Interior to monitor subsequent drilling operations.

In spite of these warning signs from the legislative branch, Watt and his successors continued to push for increased development of the Outer Continental Shelf. As a result the Congressional moratoria became an annual event, and gradually the prohibited area grew in size and geographic distribution to the point that practically the entire Outer Continental Shelf was eventually closed to the agency created to exploit it.

In the spring of 1989 George Bush inherited the Outer Continental Shelf gridlock generated by his past four predecessors. There had been only two Outer Continental Shelf sales outside of the Gulf of Mexico in the past five years in the Beaufort and Chukchi Seas off Alaska, and Congress showed no inclination to relinquish its control over the leasing process.

The controversy at that time centered on three proposed sales in California and Florida. In Florida the sale of leases below the 26th parallel had been temporarily halted when Governor Bob Martinez first sued and then settled with Secretary of Interior Hodel (see chapter 7). The compromise reached allowed a portion of the proposed sale in the eastern Gulf to go forth, but postponed sales of leases below the 26th parallel. The second half of that sale (116) was again coming up. As noted before, Florida, under a succession of governors, had opposed development in this area, and this time was no different. The state was continuing to fight exploratory drilling on tracts that had been leased under previous sales, and a number of state and local organizations (particularly in south Florida) had focused their efforts on stopping the sale.

In California, two sales were proposed. Sale 95, off of southern California, was in the area where all of the Outer Continental Shelf production on the Outer Continental Shelf outside of the Gulf of Mexico had already taken place. Sale 91, in contrast, was proposed for northern California, a "frontier" region. While opposition to both sales was high, the opposition to the southern sale paled in comparison to that for sale 91. The hearing for sale 91 was held in Fort Bragg, California, on February 3 and 4, 1988 (Freudenburg and Gramling 1994a), was attended by thousands of opponents, and provided a spectacle of opposition the likes of which have seldom been seen.

> The first speaker, the California Lieutenant Governor, left no doubt about his opposition, and the reaction of the crowd left little doubt that they agreed. The second speaker, the Chair of the Coastal Commission, spoke of his Commission's responsibility to protect the coast, and he made it clear that he would not take that responsibility lightly. The California Attorney General said not only that was he unalterably opposed to the Federal proposals, but that he considered them to be illegal; he promised that if the sale were to go forward, he would take the agency to court, and that he would prevail. A member of the local County Board of Supervisors, noting that the Board had voted unanimously to oppose the sale, was more than willing to joint in: We will fight you page by page through this nefarious document. A Supervisor from Sonoma County underlined the point: Welcome. You are surrounded. We ask your unconditional surrender. The idea of offshore drilling along this coast, he announced, was as ridiculous as paving the Grand Canyon.
>
> Yet that was only the beginning. The officials were to hear similar messages from representatives for both U.S. Senators from the state, for 23 Members of Congress, and for a string of cities and counties, all along the coast (Freudenburg and Gramling 1994a: 5)

Nor was the testimony restrained in its attempt to convey opposition to the proposed sale:

> Some of the voices thundered, others whispered, and still others choked with emotion. Guitarists performed songs that they had composed for the occasion; school children sang music and performed plays that they had written as well. At one point in the hearings, a man dressed as Coyote came up to the podium.

He noted with approval the representation of his friend the buffalo on the Department of Interior seal, but then asked melodramatically, *Where* has the buffalo gone? Are you *protecting* the buffalo? ... Coyote's worried! At another point, in the effort to demonstrate what a California-style earthquake could do to an offshore oil platform, a belly dancer, suspended horizontally by a set of accomplices, placed a plastic rig on her mid-section and began to gyrate. The oil rig performed its part of the drama impeccably, and the crowd roared its approval (Freudenburg and Gramling 1994a: 4)

The hearings became a national media event and must have left an impression on a then-campaigning Vice-President George Bush.

PRESIDENTIAL INQUIRY

One of the first actions of President George Bush, early in 1989, was to announce that he was postponing the second half of lease sale 116 off Florida and the two sales off California. The question of whether the sales should go forth was submitted to a cabinet-level Presidential task force, and the National Academy of Sciences was asked to evaluate the adequacy of the available information (a key point for the opposition to all three sales), concerning the potential impacts of the sales.

The task force met throughout 1989, and several of the public hearings were reminiscent of Fort Bragg. At the public hearings in Florida, similar creativity was used to convey the message. In north Florida, participants presented the task force representatives with bags of tar picked up off the beach, and in south Florida, one of the participants threatened to pour oil on the task force representatives, several of whom were visibly shaken. In California, the Mendocino County (location of Fort Bragg) Board of Commissioners voted not to allow the task force to meet in the county. How they would have prevented this is not clear, but the task force got the message and met elsewhere.

The "socioeconomic" panel of the National Academy committee did meet in Fort Bragg in the summer of 1989. Initially the committee's inquiries were going to be low-key, but when word of the meeting leaked out, the committee was left with little choice but to hold public meetings, where they were treated to a somewhat more restrained but still impassioned opposition to the lease sale. One of the local speakers, who made her living painting coastal scenes, broke into tears in trying to describe the impact that the development of offshore oil would have on her, and another asked if the federal government had figured the cost of suppressing armed rebellion into its cost-benefit analysis of the sale. Another, stated quite matter-of-factly, yet with obvious threatening intention, that he would not "live to see offshore development" in the area. The committee was told repeatedly that whatever means necessary would be used to stop the sale.

By late 1989 the National Research Council (the working arm of the National Academy of Sciences) released its report, finding that the available information for informed decision making was inadequate for all three sales (National Research Council 1989). In January of the following year the task force presented its findings to the president. While the report offered a number of options, none of them included going ahead as planned.[3] Several weeks later in a statement released on June 26, 1990, President Bush canceled the three sales and imposed a presidential moratorium on lease sales in these three areas and on much of the East and West coasts of the United States until the year 2000. There were no lease sales outside of the Gulf of Mexico during Bush's tenure in office.

Also later in 1990, Congress reauthorized the Coastal Zone Management Act. Amendments were added to the act stating that:

> Each Federal agency activity within or outside the coastal zone that affects any land or water use or natural resource of the coastal zone shall be carried out in a manner which is consistent to the maximum extent practicable with the enforceable policies of approved State management programs. (PL 101–508, 6208 (c) (1) (A)).

This specifically overturned the *Interior* v. *California* Supreme Court decision (464 U.S. 312 (1984)) that lease sales cannot cause impacts and thus are not subject to consistency review.

> The conferees principal objective in amending this section is to overturn the decision of the Supreme Court in *Secretary of*

the Interior v. *California* . . . and to make clear that outer Continental Shelf oil and gas lease sales are subject to the requirements of section 307 (c) (1). (House Conference Report No. 101-964, 970, October 27, 1990 accompanying P.L. 101-964, as reported in U.S. Congressional and Administrative News 2017 No. 10c January, 1991)

LOUISIANA GETS INTO THE ACT

Bush's decision and the reauthorization of the Coastal Zone Management Act were not lost on officials in the state of Louisiana. For many years, under a succession of governors, the state had objected to area-wide leasing and proposed various forms of sharing the revenue from Outer Continental Shelf development to the Department of Interior, noting the cost to the state of maintaining the physical infrastructure to support Outer Continental Shelf activities. As of 1990 the only success had been the "8g" settlements where, as the result of an additional amendment to the Outer Continental Shelf Lands Act in 1986 (P.L. 99-272), coastal states were given 27 percent of the revenue generated from lands three to six miles offshore on the assumption that some of the reservoirs under those lands could extend shoreward into state waters (see Farrow 1990 for further discussion). Midway through his first term in office, Governor Buddy Roemer began again to push for additional Outer Continental Shelf revenue sharing and a stop to area-wide leasing, meeting with Secretary of Interior Lujan in the spring of 1991.

Receiving no satisfaction from Interior, Roemer decided to use the new leverage provided by the reauthorization of the Coastal Zone Management Act and in May of 1991 directed the Secretary of the Louisiana Department of Natural Resources, Ron Gomez, to challenge Minerals Management Service's consistency determination for a lease sale (sale 135) in the western Gulf of Mexico. Under the Coastal Zone Management Act, a federal agency must certify that its actions are consistent with the state's coastal zone program and submit that review to the state for concurrence. Because the state of Louisiana had seldom objected to federal Outer Continental Shelf activities, the exception being area-wide leasing, the consistency determinations by Minerals Management Service were relatively perfunctory.

Gomez denied concurrence, the Department of Interior reiterated its position by proposing to hold the sale over the objections of the state, and in August of 1991 Roemer went to court seeking an injunction to stop the sale. Legally the state made three basic arguments: the consistency determination by Minerals Management Service was inadequate, the proposed sale was not consistent with the state coastal zone management plan, and the environmental impact statement as required under NEPA was inadequate. Conceptually, the state also made three arguments. First, the state cited Outer Continental Shelf activities as a contributing factor in coastal land loss, primarily through erosion caused by pipeline corridors and vessel traffic. Second, the state maintained that the impacts on human populations in the coastal areas contiguous to Outer Continental Shelf activities had been virtually ignored by Minerals Management Service. Third, the State objected to the process of area-wide leasing, noting:

> The State has repeatedly recommended that feasible and practical alternatives to area-wide leasing be implemented: A more carefully controlled Lease Sale (such as those successfully conducted between 1954 and 1983 as noted previously) is a feasible and practical alternative to the proposed one and will assist the proposed activity to conform to the LCRP[4] guidelines to the maximum extent practicable.
>
> This strategy would also have the benefit of mitigating:
> - Significant Contributions to the decline in value of offshore leases, and as a result, the virtual give away proposed minimum acceptable bid of $25 per acre . . . ;
> - Massive cyclical employment . . . due to unplanned and unsustainable development of the on-shore infrastructure;
> - The exacerbation of cumulative impacts which have been responsible for the gradual alteration of the physical, social, and economic environment in coastal Louisiana;

- A more rapid depletion of the OCS oil and gas reserves upon which our coastal economy has become (as a direct result of OCS leasing practice) so dependent, and consequently a shorter time for the State to diversify its economy. (Louisiana Department of Natural Resources 1991: 36–37)

While the state failed to get an injunction, the late date at which the state asked for one (several weeks before the sale) and the failure of the state to object at the appropriate stages in the lease preparation process were major factors noted in the decision.

The evidence before the secretary, *at the time the determination of consistency was made,* was sufficient for him to conclude that the lease-sale was consistent with the state management scheme. Had the plaintiffs presented the evidence they have presented to this Court to the secretary before the decision was made, perhaps his decision (or my decision) would be different, but based on the record as it was then constituted the action of the secretary was not unreasonable (Beer 1991, emphasis in original)

While the State of Louisiana was preparing systematic objections to the next lease sale, Roemer lost his reelection bid to former Governor Edwin Edwards, who promptly dropped the objections.

INTERIOR DELUSIONS

In some ways, by the time Minerals Management Service was created by Watt, the task that the new agency was handed was already out of its hands. Certainly, less than with many agencies, even within Interior, Minerals Management Service had from its inception less ability to make policy decisions concerning its supposed domain. In fact, starting with Nixon, policy decisions concerning Outer Continental Shelf exploitation were not even nominally set within Interior. At the other end, in response to perceived insensitivity to Congressional intent, Congress was increasingly closing options available to the agency to fulfill goals set at the highest levels of the executive branch. Thus, policy goals were being set outside of the agency, and the ability to carry out these goals also increasingly came to be determined outside the agency. What Interior and Minerals Management Service did control, however, in this period following the 1973–1974 embargo, was the way in which policy was implemented, and this appears to be at least part of the problem.

The process did not start with Watt, as Engler noted:

> To a disturbing extent Interior's performance has been most consistently shaped over the years by the assumption that it is holding public lands (up to one-third of the land area of the United States and the continental shelf) for disposal to private claimants. It has found itself uncomfortable in dealing with broad questions of public policy for energy. And conservation has often been, both historically and at present, an after-thought.
>
> Interior's working perspective thus has kept it a partner in development, responsive primarily to the needs of the private government which controls energy. It is an ineffectual guardian of the public domain and insensitive to the long-run public interest in the development, use, and conservation of energy resources. In many respects it serves as the first line of defense of private industry, and its personnel act as the "fifth column" within public government for keeping prices up, profits secure, and private controls of energy insulated from public accountability. (Engler 1977: 146–147)

Watt's dismissal of public concern certainly exacerbated the problem, and throughout the 1980s, Interior came to be in an increasingly confrontational posture with coastal states and communities outside of the central and western Gulf of Mexico. With Watt's departure, however, the department did not fundamentally change its approach, and by the mid-1980s, with the exception of the central and western Gulf, the Outer Continental Shelf leasing program was essentially dead in the water. The question that arises from the examination of this process is why Interior persisted in this seemingly self-destructive behavior, as the forces arranged against the program grew and the warning signs emerged in Congress.

There are a number of factors that appear to have affected the ability of those within Interior, who were at a level to even try to effect policy implementing the Outer Continental Shelf program, to see that there was a problem, let alone to find a solution. First, as Cicin-Sain and Knecht (1987; Cicin-Sain 1986) have noted, the very structure of state and federal authority on the seas is inherently problematic. In most instances the duties and functions of state and federal governments overlap or are related within the same geographic area. Within the marine context, however, there is a clear and arbitrary dividing line (three miles offshore in most cases) between state and federal authority. This line is invisible to the ecological and economic processes that constitute and use the marine environment.

> The mismatch between highly interdependent and interconnected ocean resources and processes and the nature of state and Federal authority in these areas—exclusive and geographically separate—has spawned a variety of recurrent intergovernmental controversies. . . . These revolve around: (1) questions of ownership—who shall own the resources, (2) questions of management control—who should have management authority and for what purposes, (3) questions related to spillover consequences—what side effects occur for adjacent jurisdictions and how should they be dealt with, and (4) questions regarding the distribution of costs and benefits—who shall bear the costs and who shall reap the benefits? (Cicin-Sain and Knecht 1987: 151)

A second basic factor in the inability of Interior to address the problem is that the Outer Continental Shelf leasing program has been very profitable for Interior and the federal government. To date, over $100 billion has been collected from Outer Continental Shelf lease sales and royalties. Approximately $82 billion has gone into the general fund, while $13 billion and $2 billion, respectively, have gone into the Land and Water Conservation Fund and the Historic Preservation Fund, both administered by Interior (*MMS Today* 1992).

Third, further complicating the ability of all sides to negotiate a settlement over the respective issues raised is the inability of either side to predict the outcomes of the Outer Continental Shelf leasing and development process. As noted earlier, the process is no longer running under the rules laid out by Congress or indeed any other set of predictable practices. Watt clearly circumvented the intentions of Congress by going to area-wide leasing against the virtual unanimous opposition of the states, in spite of the Congressional language to "assure that States, and through States, local governments, which are directly affected by exploration, development, and production of oil and natural gas are provided an opportunity to participate in policy and planning decisions relating to management of the resources of the Outer Continental Shelf" (43 U.S.C. § 1801).

When challenged in court, the courts ruled narrowly, leading many to conclude that they sidestepped their broader obligations (Fitzgerald 1987). Following this development, Congress circumvented the established procedures and shut down the decision-making process within Interior through the budgetary back door. As a result of all of this, even if the conflicting factions were willing to sit down and negotiate a settlement, there is a distinct possibility that no one involved in the discussions could, with good conscience, guarantee that the agreements would be carried out!

Fourth, one of the more basic reasons for the inability of those agency personnel with some control over policy implementation to anticipate the increasing problems can be explained as the age-old problem of communicating across cultural lines, lines associated with the regions in question and those associated with what has been called the culture of the agency. These are discussed in detail in chapter 7.

Fifth, and perhaps *most* important, was the ideological differences between the factions that led to the enactment of the Outer Continental Shelf Lands Act Amendments and those left to implement it. As noted earlier, the OCSLAA were passed in 1978 in response to states' opposition to the federal leasing program and the perception of Congress that the program was a closed deal between the secretary of Interior and the major oil companies. The intent of the amendments was to allow more state and regional inputs into the decision process and to more carefully account for and monitor the effects of the program (see Appendix). However, in 1980 the management of the program fell into the hands of the Reagan administration, as did the job of implementing the requirements of the OCSLAA (Freudenburg and Gramling 1994b).

The ideology of the Reagan administration was unabashedly pro-business and anti-government regulation (Barnett 1989; Riposa 1989). Although pressure from the states that opposed the program, as it had been run following the oil embargo, led to what they considered an important legal-policy victory, the implementation of the statute was left to an administration that basically disagreed with the premises of the legislation (see Barnett 1989 for a similar case) and to the not-so-subtle Secretary of Interior James Watt. The net result of that implementation was what Freudenburg and Gramling (1994b: 509) call "bureaucratic slippage," or the process by which "broad policy statements . . . [are] successively reinterpreted, both over time and across multiple layers of regulatory implementation [much like] . . . the childhood game in which a 'secret' is whispered to one person, who then whispers it to the next, and so on; the eventual secret, or the eventual implementation of the policy, can prove to have very little resemblance to the statement that started the process." The broad requirements of the OCSLAA became reinterpreted into narrow agency missions, which supported offshore development, and even narrower funding criteria for research to accomplish the required assessment and monitoring of the effects of this offshore leasing, a process that essentially put blinders on the agency (Freudenburg and Gramling 1994b). Thus, the ideological underpinnings of the Reagan administration, more particularly as implemented by James Watt, brought the federal offshore leasing program increasingly into a confrontational posture with many of the coastal states. The crown jewel of that confrontation was area-wide leasing, and in the Gulf the major opponent was Florida.

Chapter 7

The Florida Conflict

OFFSHORE ACTIVITIES IN FLORIDA WATERS

State Waters

L ike California, Texas, and Louisiana, prior to the settlement of the Tidelands issue, Florida also sold petroleum leases off of the coast. In 1941 the state entered into an agreement with Arnold Oil for exploration of coastal and inland waters. In the late 1940s Coastal Petroleum Company succeeded Arnold and obtained three leases in Florida waters. Two of these leases covered state lands from the shore to three marine leagues offshore[1] and along the coast from Naples to Apalachicola (the vast majority of Florida's west coast), and the submerged land of rivers and lakes that drain into this area. The third lease covered Lake Okeechobee (Lloyd and Ragland 1991). Initially these leases had no expiration date. Additional leases were also let by the state, and between 1947 and 1983 a total of nineteen wells were drilled in Florida state waters, including seven in the Florida Keys and two in the Pensacola Bay area, under leases that have since expired. Only one of these wells, drilled northeast of the Marquesas Keys, had a significant show of oil (Lloyd and Ragland 1991). In 1976 the Coastal Petroleum leases were renegotiated to include only the area from three marine leagues offshore, extending three miles landward, and to expire in 2017.

With growing public pressure to curtail offshore oil and gas activities, in 1989 and again in 1990, the Florida legislature effectively prohibited the sale of new leases and drilling in state waters. The exceptions to this prohibition are the remaining three leases held by Coastal Petroleum, though permits for new exploratory drilling have been denied, and there are currently serious legal problems surrounding Coastal's leases.

Federal Waters

The first federal lease sale off Florida was held in 1959, six years after the passage of the Outer Continental Shelf Lands Act and after the contentious battles with Louisiana over the state-federal boundary had been settled (chapter 2). The sale was held in what is now considered by Minerals Management Service, the Straits of Florida, Atlantic planning area, south of the Florida Keys. Twenty-three tracts were leased, and three exploratory wells were drilled south of the Marquesas Keys, as a result of that sale. All wells had been plugged and abandoned by 1961 (Lloyd and Ragland 1991).

By the 1970s industry began to show an interest in federal waters off the Florida Panhandle, and in the first sale (73) on what is now considered the eastern Gulf of Mexico planning area for Minerals Management Service, it demonstrated that collective interest with record bids. Destin Dome, a geological structure south of Destin, Florida, dominated that sale. A consortium made up of Mobil, Exxon, and Champlin bid $632.4 million to acquire six tracts over the dome, the total spent on the dome was $784 million, making it the world's most expensive single structure to date (Thobe 1974). Joint bids by Chevron, Amoco, and Union resulted in their leasing ten tracts, tracts that were to result in controversy a decade later.

An additional thirty-nine tracts were leased in the Gulf off Florida, in Gulf-wide sales in the middle and late 1970s. Several leases were sold off Florida's northern Atlantic coast in 1978, but these have all expired or been relinquished. Exploratory drilling in federal waters in the Gulf of Mexico began in 1974 and, with the exception of two wells off St. Petersburg and one on the Florida

Middle Grounds, was confined to the Destin Dome area throughout the 1970s.

Although Governor Askew had objected to sale 73 (Londenberg 1973) and concern was also expressed over subsequent leasing and exploratory activities, the real conflict between Florida and the Department of Interior started with the draft proposal for the first five-year plan, which was a new requirement under the Outer Continental Shelf Lands Act Amendments. Florida Governor Bob Graham objected to the proposed five-year leasing program, since it did not exclude offering tracts below the 26th parallel.[2]

Conflict with Minerals Management Service started soon after its creation and hinged on then Secretary of Interior James Watt's revised five-year leasing plan, which initiated area-wide leasing, and proposed a sale offering virtually the entire eastern Gulf of Mexico, including the area south of the 26th parallel. Again the same issue was raised with Secretary Watt, and the state, citing lack of sufficient information, recommended deletion of this area in the sales authorized under the five-year plan. However, on January 5, 1984, Minerals Management Service did hold an area-wide sale for the eastern Gulf of Mexico (sale 79) with tracts offered south of 26 degrees north latitude. Sixty tracts were leased. The tracts were southwest of Naples in an area designated as Pulley Ridge, after a subsurface geological formation. With the sale of these tracts, Watt really ignited the controversy between the Department of Interior (Minerals Management Service) and the state of Florida, a controversy that has burned for over a decade. The following year, Minerals Management Service again held an area-wide sale (sale 94) in the eastern Gulf of Mexico, again over the objections of the state of Florida, and thirteen more tracts were leased in the Pulley Ridge area.

As the proposed five-year plan for 1987–1992 was being developed and finalized, the issue arose again. Again citing lack of information (Johnson and Tucker 1987) Graham's successor Governor Bob Martinez, recommended to Watt's successor, Secretary of Interior Donald Hodel, a number of deletions, including the area south of 26 degrees north latitude. In addition, on May 5, 1987, Governor Martinez released the draft report of an independent scientific review of Minerals Management Service's studies of the Southwest Florida Shelf. While the review found that "The studies have added a significant body of information regarding the biology and physical oceanography of the shelf off southwest Florida," it concluded that "The studies did not provide the type of information nor detail to determine the specific impacts of oil and gas activities on certain sensitive habitats off southwest Florida" (Governor's Office of Planning and Budget 1987: 30). This issue escalated into a full-scale political battle between Martinez and Hodel when the third area-wide lease sale (sale 116) proposed to open the same area. As a result of a suit by Martinez and a compromise between the two, the southern portion of proposed sale 116 (south of the 26th parallel) was delayed. Following the examination of the issue by a Bush-appointed presidential task force and the National Academy of Sciences, leasing in the area has been postponed until after the year 2000 (see chapter 6).

The fire has been kept alive, however, by proposals for exploratory drilling by Union Oil on two of the Pulley Ridge blocks (629 and 630) and by Mobil on an additional block (799) (Russell 1990a, 1990b; Maloney 1990). Once again consistency under the Coastal Zone Management Act has become a key issue. In 1988 the state objected to the Department of Interior's determinations that the plans were consistent with Florida's Coastal Management Program, and Union and Mobil appealed to the Secretary of Commerce, as the Coastal Zone Management Act allows.

With the reauthorization and strengthening of the Act by Congress in the fall of 1990 and the implementation of a new state policy (to oppose all drilling within 100 miles of the Florida coast), Florida also objected to a consistency determination allowing Chevron to continue exploratory drilling aimed at delineating natural gas resources off western Florida in the Destin Dome area (Smith 1992). Chevron had quite openly planned for eventual development of these gas resources if future exploratory drilling met expectations, and this development would represent the first on the Outer Continental Shelf off Florida. In January of 1993 the Secretary of Commerce upheld Florida's objection to the two proposals off south Florida but failed to uphold the one off west Florida. The unique environment of south Florida and the probability that the resource off west Florida was natural gas (hence less danger of an oil spill) were major factors in the different results.

The impetus for the objections and the hardening of the state's position during the 1980s and early 1990s was Watt's initial area-wide sale off south Florida. The explanation of why south Florida in the mid-1980s was a different ballgame than Florida in the 1940s and 1950s,[3] the western part of the same state, or other Gulf coast states, and why those in the industry and Interior did not realize this, lies in the history and characteristics of the regions and in the emergence of unique cultures,[4] both in the agency and industry, and in south Florida.

INDUSTRY AND AGENCY CULTURE

The culture of the offshore industry and the agency was shaped by the historical and geographic origins of Outer Continental Shelf activities, which flourished in the supportive atmosphere of the central and western Gulf of Mexico. That process has been detailed in the earlier chapters of this volume, but a summary will be useful before examining the emergence of south Florida culture and comparing it to the process in Louisiana. As will be recalled, all of the production on the Outer Continental Shelf between 1954 and 1963 was in the central and western Gulf. In 1959 there were lease sales off the Florida Keys, and in 1963 and 1964 there were sales in the Pacific, off central California and the Washington-Oregon coasts. Exploration in these areas failed to find commercially valuable quantities of oil, and the leases expired (Gould et al. 1991). It was not until 1966 that the first lease sale was held off southern California, and it was less than three years later, in early 1969, when Santa Barbara experienced a major spill of Outer Continental Shelf oil.

Following the spill, no leases were held outside of the central and western Gulf for another five years, until the OPEC embargo led to calls for further leasing from the federal government and the oil companies, although not from the residents of the potentially affected regions. In the Gulf, the agency and industry had been developing a close working relationship over two decades, coming in the process to share a perspective that remained virtually unchallenged until the mid-1970s.

In a systematic comparison of the factors that have led to the acceptance of offshore activity in Louisiana and its rejection in California, Freudenburg and Gramling (1993, 1994a) note a variety of socioenvironmental factors that led to the respective positions. Included in these are a number of historical factors and the economic overadaptation noted in chapter 5. In addition, they note a number of biophysical factors that have influenced the respective positions such as the awe-inspiring Pacific coast and the less impressive Gulf; the accessibility of the California coast and the lack of access in Louisiana; the broad continental shelf in the Gulf, which minimizes user conflicts as opposed to the narrow one in California, which maximizes such conflicts; and the numerous waterways and harbors in Louisiana, as opposed to the comparative scarcity of navigable waterways and harbors in California.

The point is that southern Louisiana and northern California have developed different and very distinctive ways of viewing the world, particularly in relationship to the question of offshore oil and gas development. In fact, the very words "offshore oil development" will evoke different emotions, reactions, and responses and will elicit different assessments of the motives of the individuals who speak the words. Just as states or regions can develop distinct subcultures, so too can agencies. There are even advantages to agencies having a distinct coherent view of the world; perhaps most importantly, it ensures that agency personnel interact consistently with the outside world, based on the shared norms of the distinct agency subculture (Peters 1978; Smircich 1983). However, as the National Research Council pointed out with respect to Minerals Management Service, it is important to remember that this is *a* distinct coherent view of the world, not *the* view of the world.

> The Minerals Management Service must recognize that government officials have viewpoints—sometimes many viewpoints—and that there is no particular reason to expect the government view to be understood or even widely shared by others—especially if the others are far removed or culturally diverse. In other words, the officials' understandings and definitions of OCS issues and problems—and solutions to them—are as much based on those persons' perceptions, biases, culture, and experience as are those of any other affected person or community. (National Research Council 1992: 68)

As is clear from the preceding portions of this volume, the story of the actual development of offshore oil and gas in the United States is primarily a Gulf of Mexico story. For many years the Gulf of Mexico was the only offshore game, and personnel in federal agencies involved with the program most likely gained their experience in the Gulf of Mexico and accepted the Gulf model of development; a model that has failed outside of the distinct environment in which it arose (Freudenburg and Gramling 1994a).

SOUTH FLORIDA

Although Florida was "discovered" by European Explorers before Louisiana and settled by the Spanish before French settlement occurred in Louisiana, for most of their respective histories since European discovery, southern Louisiana has had a much denser population.[5] As settlement began to occur in the two coastal regions, the nature and location of their extensive wetlands became a factor. Both southern Louisiana and southern Florida are characterized by extensive coastal wetlands, which have a number of similarities, not the least of which is a mosquito population density that would awe the uninitiated. The coastal marsh in Louisiana has been described earlier. Having been deposited by the wandering Mississippi River, much of the Louisiana "coast" consists of wetlands, which, like the southeastern tip of Florida, meld gradually into the Gulf of Mexico. However, there are differences between the regions also. The Louisiana deltaic plain is dissected by hundreds of rivers and bayous that flow southward toward the coast, and it is along the natural levees of these waterways that habitable land elevated above the coastal wetlands is found. Accordingly, it was primarily along these natural levees that European settlement spread, spurred in the mid-1700s by French-speaking exiles from British Canada. The long "string towns" (Kniffen 1968) that emerged, centered on subsistence agriculture and an annual harvest of the many renewable resources of the surrounding marsh and swamp (Comeaux 1972; Davis 1990). This led to an extractive orientation toward the coastal wetlands. Follow ing the Louisiana Purchase in 1803, more affluent American settlers bought up much of the more fertile land and established plantation agriculture, centered on sugar cane rather than cotton (Sitterson 1973). In the process they pushed the original Acadian settlers further into the marsh and the natural levees of smaller streams, raising the population density in these areas. Throughout it all, water transportation tied communities together and to their markets.

Southern Florida is characterized by an older geological foundation and a limestone and sand, rather than a silt, foundation. As a result much of the southern part of the state has sandy soil, which is marginal for agricultural purposes. Its wetlands are almost as extensive as those of coastal Louisiana, but evidence a different hydrological regime. Much of the central peninsula of Florida is shaped like a huge shallow bowl. The Kissimmee River drains this shallow bowl southward during the summer rainy season to Lake Okeechobee, an inland freshwater sea that has been slowly filling with aquatic vegetation for centuries, forming peat bogs. The further decay of these bogs produced the rich organic "muck" lands south and east of the lake, which were later to be exploited as sugar cane fields and truck farms. Because these lands were frequently under water during the summer, they were not available to the early inhabitants of south Florida, until drainage and "development" efforts began in the 1920s. The southern end of Lake Okeechobee overflows southward into a great shallow river of grass, laced with mahogany and cypress hammocks, which averages over fifty miles wide but in most places is only inches deep. As these Everglades approach the southern coast, grass, cypress, and mahogany are replaced by mangroves, and the "solid" ground dissolves into thousands of estuaries and mangrove islands.

Southeast and southwest of Lake Okeechobee, along the Atlantic and Gulf of Mexico coasts, beach dunes and loosely connected barrier islands provide land high enough for human habitation. As a result of this geomorphology, early settlement was primarily confined to the narrow coastal strips, but without rich alluvial soils like those in Louisiana, agricultural opportunities were limited. In addition the more fragile ecosystem of the Everglades provided less in the way of renewable resources for harvest. There were, of course, virtually no roads for early European settlers, and lacking Louisiana's network of inland waterways, water transportation was forced to rely more heavily on the Gulf and Atlantic.

Both of these, however, are formidable water bodies requiring seaworthy vessels and secure ports. Thus for many areas of the coast, particularly the Atlantic coast, transportation was problematic. Climatically, southern Florida is borderline tropical, while southern Louisiana is subtropical.

The spread of the railroads following the Civil War altered transportation patterns in both southern Louisiana and southern Florida, as they did with much of the country, but the alteration was much more dramatic in Florida. While transportation came to be faster and more reliable in southern Louisiana, because the rails followed the natural levees of the waterways, settlement patterns continued in the same basic structure.

Shortly before the discovery at Spindletop started the push for oil under coastal Louisiana, the railroads began a very different development scenario in south Florida. The name most closely associated with that scenario was Henry Flagler. Flagler had amassed a considerable fortune as a partner in John D. Rockefeller's Standard Oil Company, and at retirement age, Flagler moved to St. Augustine, Florida, to start a new career. The enterprise consisted of a combination of building hotels along the Florida east coast and the construction of rail lines to carry guests to them (Tebeau 1971). As a consequence of Flagler's pecuniary efforts in southern Florida, some areas of the coast, off the traditional shipping routes, became accessible by modern transportation for the first time. With the turn of the twentieth century, both coastal regions were on the verge of extensive development activities but via very different paths. Louisiana was poised to move in the direction of mining its nonrenewable resources, while southern Florida was postured to "mine" its renewable amenity-based resources.

By shortly after the turn of the century, Flagler's lines connected the east coast of Florida as far south as Homestead and his hotels could accommodate 44,000 guests. In 1905 Flagler undertook a railroad bridge to the Florida Keys, and although popularly labeled Flagler's folly, the link was finished in 1912. As the railroads moved southward, the unique climate in southern Florida began to play a role.

Southern Florida, at least on the east coast, first began its current development strategy as a place where the wealthy could escape the winter cold. John D. Rockefeller spent his last twenty winters at a Flagler hotel in Ormond, Florida, and during this time, Miami Beach was developing as "millionaire's mile," a series of winter homes for the wealthy. However, other trends were beginning that would not only make these facilities available to the less affluent but that would encourage their use.

The first major factor occurred during the early twentieth century when leisure came to be recognized not as a luxury but as a necessity and even a right. Henry Ford was one of the entrepreneurs who quickly saw the relationship between leisure and consumption and was a major proponent of a shorter work week. The five-day work week became the norm (Kaplan 1975), and hours worked per week fell from seventy in 1890 to thirty-seven by 1960 (Kaplan 1960). Along with the growing importance of the "vacation," earlier retirement age, and growing median income, this new leisure made it possible for increasing numbers of the middle class to use such facilities as Flagler's hotels.

A second trend that reinforced the first one was the technological development of transportation. While Flagler's railroads continued to be important through World War II, the growth of the automobile, coupled with the desirability of traveling for a vacation, was a key to the development in southern Florida. The highway network spread. In 1928 the Tamiami Trail was completed through the Everglades, and for the first time there was a connection between the east and west coasts of southern Florida. With Eisenhower's support of the interstate highway system in the mid-1950s, road transportation was given a major boost. Interstate 75 from Detroit, down the west coast of Florida, and Interstate 95, down the east coast of Florida, became major routes for the developing tourism.

A third trend that assured southern Florida's role as the tourist mecca of the nation was the change in fashion, particularly women's fashions, after World War I and continuing to date. As Steele (1985) has noted, in the 1920s, along with an accent on the youthful look, and a blurring of the distinction between the dress of women of different social classes, a major trend in women's clothing was the display of more and more of the body. By the end of World War II, bathing suits that displayed the shape and much of the surface of the body were fully accepted.

A final important trend for the future of south Florida was the emergence of the suntan as a status symbol. Throughout most of U.S. history, a tan was a mark of the working class, indicating out-doors work, and the upper and middle class, particularly women, avoided exposure to the sun (Lurie 1981). Supposedly, the suntan as a fashion was introduced from the French Riviera by Gabriel Chanel in 1920 (Lurie 1981). As the display of more and more skin became acceptable, the suntan gained status, particularly when displayed in northern climes in the winter. Going to the beach be-came a major social activity, and "working" on a tan became a way to illustrate conspicuous leisure (Veblen 1934), even for the middle class. Slowly an entire sector of the economy emerged around coastal-beach tourism. Today, not only is it possible for millions to lie under the sun happily cultivating carcinomas, but local and national norms increasingly allow them to do it over more and more of their bodies. Today, the "millionaire's mile" on Miami Beach is the home of the Fontainebleau Hilton, and the beach in front is topless.

While beach-oriented tourism became an important economic activity in many areas of the coastal United States, it is a seasonal activity in most areas and many of the businesses oriented toward this market literally have to make their annual income during the Memorial Day to Labor Day "season." This is not true in south Florida, as the unique climate provides a viable, year-round sea-son. The geomorphology allowed settlement only along the coasts, where, with the exceptions of the rich soils south and east of Lake Okeechobee,[6] the sandy soils provided few economic alternatives. Thus, by the 1950s southern Florida had found its niche, winter coastal tourism. An additional technological development, air con-ditioning, has extended this season to a year-round activity in south Florida for the last several decades, although the winter sea-son is still the most crowded one. In many coastal areas of the United States, coastal tourism is the dominant element in the economy; in southern Florida it became *the* economy. In 1990 the Florida Keys (Monroe County) had a population of 78,024 and 6.8 million visitors, who donated over $30 million in sales taxes alone (White 1993). As an indication of the level of specialization, the same year the state had 1,096,082 rental units, and 45,583 public food serving establishments with a combined seating capacity of

2,811,733 (Morris 1991). This is sufficient capacity for the populations (1990) of Maine, Vermont, and Rhode Island to eat out in
Florida in one sitting.

In addition, retirement to Florida, either full time or during the
winter (snowbirds), has become increasingly popular. This,
coupled with the growing labor demands of the service sector that
supports tourism or retirement, has led to rapid in-migration, literally building the population of south Florida since World War II.
Today an average of 1,000 people a day move to Florida, most to
the southern part of the state. Figure 7.1 plots the populations for
Louisiana's and Florida's southern coastal parishes/counties from
1850 to 1990. The trend is very clear. From before the Civil War
until World War I, south Florida had a very scarce population,
which grew slowly. The population increased between the two
world wars but skyrocketed following World War II, particularly
after 1950. In 1950 the population of the five southernmost counties in Florida (Broward, Collier, Dade, Monroe, and Palm Beach)
was just over 700,000. By 1990 the population of these same counties was almost 4.5 million, a growth rate of over 500 percent. Most
of the population of these counties is concentrated along relatively
narrow coastal strips, or, in the case of Monroe County, in the
Florida Keys.

This growth rate, geographic location, and tourist-oriented
economy have had massive effects on the subculture of south
Florida. Because of the rapid population growth due to in-migration, most of the current population comes from outside Florida,
the South, or even from outside the country and, consequently,
from a variety of cultural backgrounds. As a result, there is little
regional indigenous culture. In addition, because many of the current residents are recent arrivals in south Florida, what common
norms that may have arisen are also recent.[7] Finally the common-
alities that they do share, which shape these emerging norms, con-
sist of an unavoidable immersion in some aspect of the economic
and cultural process of beach-oriented tourism and a proximity to
the beach. Many of the residents have retired to Florida to enjoy
various elements of the amenities offered by the combination of
weather and the coast, becoming in some ways permanent tourists. Another large segment of the population works in the service
sector in jobs that do not pay that well, but they remain because

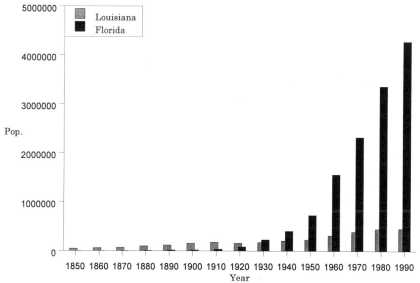

Fig. 7.1. **Population, Louisiana and Florida Southern Coastal Counties.**
Source: U.S. Department of Commerce 1850–1990.

they can also enjoy the amenities of the climate and coast. Both groups enjoy many of the same amenities as the millions of tourists who spend billions of dollars annually in south Florida. Thus, to say that southern Florida is economically dependent on coastal tourism would be an understatement, but it would still be only a part of the picture of the opposition to offshore oil.

There are two other basic factors involved. First, from an economic perspective, southern Floridians do not see that they have anything to gain from the development of offshore oil but do perceive that they have much to lose. Given the mobility of Outer Continental Shelf operations (noted in chapter 3), and from past experience, Floridians know that few, if any, jobs or other economic benefits would go to residents of the state. The rigs, support vessels, and crews would come from southern Louisiana, and local purchases would be limited to such items as potable water, which is already in short supply in south Florida, and diesel fuel.

Although sharing the revenue generated by the federal leasing program has been proposed, mentioning this in south Florida usually evokes antagonistic responses along the lines of "they can't bribe us" or "this is an insult." Furthermore, even if some form of sharing the federal revenue generated by Outer Continental Shelf production with the states were to be put in effect, this would probably be an insignificant input into the regional economy. The last proposal put forth by the Department of Interior (Minerals Management Service 1990a) was that starting with 1993 states and coastal counties within 200 miles of an Outer Continental Shelf tract would split 12.5 percent of the new federal royalties generated on that tract.

Setting aside the fact that this proposal excluded the bonuses paid for leases, which historically have been about half of Outer Continental Shelf revenues, the figures are not likely to sway Floridians. According to Minerals Management Service (1990b), the estimated petroleum resources of the entire eastern Gulf of Mexico are 1.25 billion barrel equivalents of oil. This is a measure that represents both estimated oil and gas reserves by converting gas reserves to the energy equivalent of barrels of oil. Assuming the most optimistic economic estimates that (1) the entire reserves are oil (which we know they are not, since Chevron has found significant gas reserves off western Florida); (2) oil continues to sell at the equivalent of $20 a barrel over the life of production; (3) all the reserves are produced over twenty years; and (4) Minerals Management Service gets one sixth of the revenue generated, this would amount to $26,041,600 per year.[8]

The controversial Pulley Ridge tracts comprise about one third of the currently active tracts leased in the eastern Gulf. Even if they were to produce one third of the revenue (which is not likely), this would mean about $4.3 million for the state of Florida and potentially some fraction of $4.3 million for each of seventeen counties annually (average about $250,000).[9] What Floridians, particularly south Floridians, see at risk is the multibillion dollar (annually) tourism industry. Florida has always maintained that the tourism industry is very fragile, depending heavily on reputation and the state's unique and uniquely fragile environment, and that any hint of trouble, such as an oil spill, would have devastating effects on the local economy. To a certain extent this fragility was borne out in the fall and winter of 1993, when the murder of several German tourists in Miami caused a significant impact on the winter tourist season (Applebome 1993; Fedarko 1993; Kinzer 1993).

   Yet this is not the whole issue; most Floridians do not want the jobs or other benefits of Outer Continental Shelf development, even if they could be guaranteed not to impact the environment, and opposition to Outer Continental Shelf activities, particularly in south Florida, cannot be fully understood from an economic or environmental perspective. The unique history, distribution, and recent arrival of the population in southern Florida has led to a subculture of coastal recreation, where the dominant themes are aesthetics, the coast, and casual living. As one respondent noted "what we're selling here is clear water, white sand, and peace of mind, and offshore oil isn't compatible with any of those." The distinctive nature of the culture becomes evident when one talks with residents of southern Florida about offshore oil and gas development. What becomes apparent quickly is not just opposition to the activity but a failure to understand why anyone familiar with the region would even think that offshore development was an option. In the words of one Key West resident "it's just not compatible with the area or the lifestyle down here." On the other side of the coin, of course, it is just as inconceivable to those in the industry and the agency that people and their elected representatives would be willing to allow potential petroleum resources to remain in the ground.

COMMUNICATING ACROSS BOUNDARIES

Thus, the formation of unique subcultures in the region, Minerals Management Service, and industry provide differing and frequently conflicting views of the world. When serious challenges to these views begin, the conflicts with opposing groups can lead to a closing of the ranks and further commitment within the groups to the very perspectives that created the conflicts in the first place (Coleman 1957). Because of the divergent processes over time and different socioenvironmental factors that led to the opposing worldviews, they are fundamentally different but literally socially constructed realities (Berger and Luckmann 1966).

   A factor that resists the reconciliation of these diverse realities is that the battles are increasingly being fought out in an arena that guarantees an adversarial approach to decision making—the legal system. As Gramling and Freudenburg (1992a) have noted, the litigious nature of the debate over proposed Outer Continental Shelf developments has further increased the tendencies toward intractable positions, not just among coastal regions and environmental groups but also in federal agencies, such as Minerals Management Service, and industry. Once the conflict enters the arena of litigation, two basic things happen. First, the lawyers for all parties insist the parties *not do* precisely what virtually all risk communication and conflict management literature insists that people *must do*, which is to start talking to the other side (Hance et al. 1988).

Second, as a result of this first factor, any information that does get from one side to the other is filtered through the respective side's attorneys, a process that guarantees that the information is at least third hand. Thus, (1) litigant A talks to attorney A, who (2) talks to attorney B, who (3) talks to litigant B. To further confuse the information flow, the two middlemen in the information transfer process probably have more in common (law degrees and a pecuniary interest, at a minimum) than do the two sides in the litigation or either side has with their attorney, and the two middlemen are being engaged only as long as the conflict continues. The net result is a "spiral of stereotypes" (cf. Freudenburg and Pastor 1992). Once the information loop is severed by the threat of litigation, both sides begin to talk not *to* the other side but *about* the other side, and third-hand information, rumors, and speculation become the only real sources of information about the presumed adversaries in other camps (cf. Coleman 1957). By this time, such shared invectives as "pot-smoking granola-chewing hippies" and "pot-bellied Texans in pointy-toed boots" have become part of the argot for the two camps, and the question has become not how differing approaches view the use of the coast but who is "right" or "wrong." Elevated through the political process to matters of policy, the conflict grows in intractability as "the Department of Interior" and "the state of Florida" find themselves as the primary combatants, and the question gets even more confrontational and eventually translated into who will "win" or "lose."

Epilogue

Offshore Oil at the Edge

CURRENT STANDING OF THE OUTER CONTINENTAL SHELF PROGRAM

The effect of the federal leasing program and the way in which it has been pursued over the last two decades has been to close down the Outer Continental Shelf, with the exception of the central and western Gulf of Mexico and portions of Alaska, to the very agency created to exploit it. By 1993 for all practical purposes, outside of the central and western Gulf of Mexico, the federal leasing program had stopped, and with dwindling known reserves in the Gulf (Gould et al. 1991), declines in production, lease sales, and revenues from this source has already started and will inevitably continue. The year 1992 was the first year on the Gulf of Mexico Outer Continental Shelf when the number of platforms removed exceeded the number of platforms installed, ironically providing an additional and much needed economic activity for the existing Outer Continental Shelf support infrastructure (Smith 1993). The reauthorization of the Coastal Zone Management Act in 1990 arguably provided a mechanism for blocking lease sales and other Outer Continental Shelf activities for states wishing to use it, and Florida is busily testing the limits of the consistency provisions. Given the unanticipated consequences of the policy initiatives set in motion by Project Independence, the question arises as to what the original goals of the policy were, whether they were realistic when they were initiated, are realistic now, or will be in the future?

Although energy independence was the stated goal of Nixon's initial expansion of the Outer Continental Shelf leasing program and also the stated goal of the Reagan administration's further expansion, as initiated by Watt, relatively simple math will demonstrate that this goal was never a realistic one. Table E.1 shows the estimated, undiscovered, economically recoverable oil and gas reserves on the U.S. Outer Continental Shelf. A number of conclusions can be reached by examining Table E.1. First, over two thirds (68.22%) of the estimated energy reserves on the Outer Continental Shelf lie in the central and western Gulf of Mexico. Second, if the estimated reserves of the eastern Gulf of Mexico (where production is imminent) and southern California (the only area outside the central and western Gulf where production is currently [1994] ongoing) are added in, these areas account for over 80 percent of the estimated energy reserves. Third, of the 16.68 billion barrels of oil equivalent in the central and western Gulf, 11.28 billion barrel equivalents (or 67.6%) is natural gas (Gould et al. 1991). However, according to Gould et al. (1991) half of the undiscovered reserves in the central Gulf (51%) and over a quarter of the undiscovered reserves in the western Gulf (27.7%) are estimated to be under areas already leased. This leaves an estimated 9.55 billion barrel equivalents of oil in the central and western Gulf available to the federal leasing program, a figure that could have been projected relatively accurately when Watt was making leasing decisions in the early 1980s. To put this total in perspective, in 1981 the United States was importing about 5.4 million barrels of oil a day, or a little less than 2 billion barrels a year. These imports accounted for about one third of the 16 million barrels consumed daily or 5.8 billion barrels consumed annually.[1] Thus, even if Outer Continental Shelf oil and natural gas had replaced oil imports, the estimated unleased energy available in even this most promising area of the Outer Continental Shelf leasing program would have offset less than five years of current imports and have supplied only two years of current total consumption. Adding in the remainder of the estimated Outer Continental Shelf unleased reserves would only extend these figures a year to cover imports and a fraction of a year to cover consumption.

Table E.1. Outer Continental Shelf Reserves Estimates

PRESTO[a] RANKING OF OUTER CONTI-NENTAL SHELF AREAS	ESTIMATED ACRES IN MILLIONS	MARG. PROB. OF H[b]	ESTIMATES OIL IN BBARRELS	GAS IN TCF	BBLS OF OIL EQUIVALENT
1. Central Gulf of Mexico	48	1.00	3.82	37.66	10.52
2. Western Gulf of Mexico	36	1.00	1.58	25.40	6.10
3. Southern California	30	1.00	1.31	3.01	1.84
4. Chukchi Sea, AK[c]	30	0.23	5.96	0.00	1.36
5. Eastern Gulf of Mexico	76	1.00	0.95	1.68	1.25
6. Northern California	29	0.78	0.89	2.45	1.03
7. Central California	15	0.90	0.50	0.82	0.58
8. Mid-Atlantic	82	0.44	0.22	5.35	0.52
9. Beaufort Sea, AK[c]	52	0.23	1.66	0.00	0.38
10. South Atlantic	106	0.23	0.21	4.60	0.24
11. North Atlantic	51	0.39	0.11	2.54	0.22
12. Washington-Oregon	48	0.25	0.19	1.97	0.14
13. Straits of Florida	11	0.19	0.34	0.57	0.08
14. Navarin Basin, AK[c]	37	0.03	1.14	0.00	0.03
15. Gulf of Alaska[c]	132	0.04	0.98	0.00	0.04
16. St. George Basin, AK[c]	70	0.02	0.39	0.00	0.01
17. N. Aleutian Basin, AK[c]	33	0.02	0.61	0.00	0.01
18. Kodiak, AK[c]	89	0.03	0.43	0.00	0.01
19. Norton Basin, AK[c]	25	<0.01	0.58	0.00	neg.
20. Cook Inlet, AK[c]	5	<0.01	0.17	0.00	neg.
21. Hope Basin, AK[c]	12	<0.01	0.50	0.00	neg.
22. Shumagin, AK[c]	83	0.01	0.28	0.00	neg.
23. Aleutian Arc, AK[c]	161	neg.	neg.	neg.	neg.
24. Aleutian Basin, AK[c]	47	neg.	neg.	neg.	neg.
25. Bowers Basin, AK[c]	90	neg.	neg.	neg.	neg.
26. St. Matthew-Hall, AK[c]	54	neg.	neg.	neg.	neg.

Source: Minerals Management Service 1990b; Americal Petroleum Institute 1993.
(a) Probalistic Resource Estimates Offshore. (b) Marginal probability of hydrocarbons (H), given here as a decimal fraction of 1.00 as the certainty, expresses the chance of occurrence of oil and gas in commercial volumes, using existing technology and under current economic conditions. (c) Gas in Alaska was assumed to be noneconomic as of 1/1/90. The conditional estimates represent the possible range of undiscovered resources present given the conditions that hydrocarbons exist in the area of study. Risked resources incorporate the probability that an area is devoid of commercial hydrocarbons. Oil is given in billions of barrels and gas in trillions of cubic feet. Conditional gas estimates are converted to energy equivalent barrels of oil BOE value, which is added to the liquid volume for a total BOE figure. The conditional mean is multiplied by the marginal probability to yield the risked total in billions of BOE. Areas considered to have little or no potential for commercial development are designated as negligible (neg.).

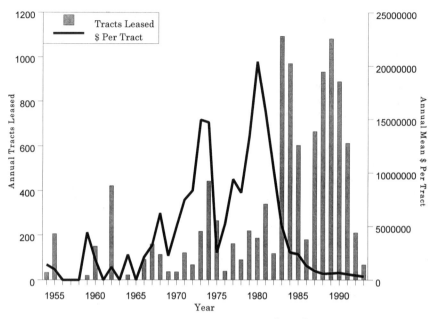

Fig. E.1. Bonuses and Tracts Leased. *Source*: Minerals Management Service, 1993.

Given that these figures were presumably known, or at least knowable, by Watt at the time he was pushing area-wide leasing, one has to wonder at the motivation(s) behind this policy decision. There are those who have suggested that the motivation was financial. Given rising budget deficits during the Reagan administration, the temptation was certainly there, and at least a superficial analysis of the data could lead one to believe that such a policy would be successful. Fig. E.1 charts the mean annual value of tracts leased by the Department of Interior against the number of Outer Continental Shelf tracts leased annually in the Gulf between 1954 and 1993. Examining the figure, it certainly appears that increased lease sales associated with Project Independence in 1974 and again during the late 1970s and early 1980s led to disproportional increases in revenue. However, when area-wide leasing was initiated in 1983, it was a totally different ball game. Until area-wide leasing, the largest number of tracts to be offered in a given sale was 515 (sale 35, in 1975), the first area-wide sale (sale 72 in 1983) offered 7,050 tracts, a 1,268 percent increase over the largest

Table E.2. Tracts Leased By Region, 1954–1992

YEAR	GULF	ALASKA	ATLANTIC	PACIFIC	TOTAL
1954	33				33
1955	206				206
1959	20		23		43
1960	150				150
1961	2				2
1962	422				422
1963				57	57
1964	22			101	123
1965	1				1
1966	91			1	92
1967	160				160
1968	113			71	184
1969	37				37
1970	36				36
1971	121				121
1972	68				68
1973	217				217
1974	443				443
1975	266			56	322
1976	39	76	93		208
1977	162	87			249
1978	90		43		133
1979	220	24	102	54	400
1980	187	35			222
1981	340	14	98	60	512
1982	118	121	26	39	304
1983	11	155		8	174

Beginning of Area-Wide Leasing, May 1983

YEAR	GULF	ALASKA	ATLANTIC	PACIFIC	TOTAL
1983	1,082		48		1,130
1984	971	390		23	1,384
1985	604				604
1986	180				180
1987	665				665
1988	934	564			1,498
1989	1,082				1,082
1990	891				891
1991	614				614
1992	211				211
Total	10,809	1,466	433	470	13,178

Source: Gould et al. 1991, Minerals Management Service 1993.

offering to that date. These increased offerings found ready buyers. Over 60 percent of all leases sold on the Outer Continental Shelf since the initiation of the federal program in 1954 have been sold in the decade since area-wide leasing was initiated, and this is in spite of the fact that most of the Outer Continental Shelf outside of the central and western Gulf of Mexico, has been closed by Congressional moratoria since area-wide leasing went into effect (Table E.2)!

The effect of the offshore glut in the Gulf was inevitable. As Fig. E.1 shows, following a nosedive by both sales and bonuses as a result of the crash of the world oil market in late 1985, the sale of leases in the Gulf rose again to new highs in the late 1980s. Unfortunately for Interior, bonuses did not show a similar trend. By 1985, two years after the initiation of area-wide leasing, the average price of an offshore tract had fallen to a new low (Fig. E.1). While the 1985 and 1986 results could be attributed to the fall of the crude oil commodities market during this period, by 1987 crude oil prices and the purchase of federal offshore lands was rising steadily, but the value of those lands was not. From 1987 through 1992 (the period of some of the lowest dollar per tract prices in the history of the federal offshore leasing program), over a third (38%) of all federal offshore lands sold since the initiation of the federal program were turned over to private investors at bargain basement prices.

Although as late as January of 1993, Scott Sewell, then director of Minerals Management Service, was espousing "a goal of reducing oil imports" through Outer Continental Shelf leasing (*MMS Today* 1993: 7), barring some massive realignment of the world energy playing field, this is simply not realistic. With the vast majority of the Gulf's reserves depleted, under production, or leased; with most of the leases sold over the last decade; and with all unleased tracts on the block at each sale in the central and western Gulf, production on the Outer Continental Shelf has continued to fall.

A simple look at the map produced periodically by Minerals Management Service, depicting the historic leasing patterns (Minerals Management Service 1989), will quickly put things in perspective. Inside the 400m contour (1,300 feet), the overwhelming majority of the tracts in the central and western Gulf are either currently leased or were leased in the past and have expired or been relinquished. Thus, the remaining potential lies in deeper and deeper waters of the Gulf, which are increasingly marginal from a financial perspective and hence increasingly less likely to be developed. Since the majority of the estimated Outer Continental Shelf reserves are located in the Gulf, this brings the entire Outer Continental Shelf leasing program into question. What is its future, or does it indeed have one?

OFFSHORE OIL ON THE EDGE

As of this writing (1994), there have been no Outer Continental Shelf lease sales outside of the central and western Gulf of Mexico in five years and only two sales outside the Gulf, both in Alaska, in the last decade. Furthermore, there appears to be little chance that Congressional or presidential moratoria will be lifted to change this picture. What then are the pivotal points on which the Outer Continental Shelf leasing program hinges and what are the issues that Interior will have to address in order to bring some form of realistic national program about? Almost two decades of familiarity with the Outer Continental Shelf program[2] leads this writer to the conclusion that there are five major issues, three specific and two very general.

1. The Alteration of the Environment in the Gulf. The aggressive Outer Continental Shelf leasing program in the Gulf of Mexico has resulted in fundamental changes in the human and physical environments along coasts of Louisiana and to a lesser extent Texas. The single *largest* impact of the Outer Continental Shelf leasing program has been in the Gulf of Mexico and on the human environment, but the relationship between these impacts and an aggressive federal leasing program has not been acknowledged by Interior, nor under the current leasing scenario can it really be addressed. The bottom line is that the Outer Continental Shelf Lands Act, as amended in 1978, gives the Department of Interior broad mandates and powers to manage and balance the impact of the federal Outer Continental Shelf program against development activities. However, the current area-wide leasing strategy essentially abrogates this authority, leaving the question of scope and range of federal sales and consequently the impact of those sales to the buyers.

2. The Focus of the Environmental Studies Program. The Outer Continental Shelf Lands Act Amendments require the Secretary of Interior to assess the impacts of Outer Continental Shelf activities and to "monitor the human, marine, and coastal environments of such area or region in a manner designed to provide time-series and data trend information which can be used for comparison with any previously collected data for the purpose of identifying any significant changes in the quality and productivity of such environments, for establishing trends in the areas studied and monitored, and for designing experiments to identify the causes of such changes." (43 U.S.C. § 1346 Sec 20 (a)). In spite of the massive impacts on the human environment in the Gulf of Mexico, this is the area that has received the *least* attention from Minerals Management Service in terms of assessment and monitoring (National Research Council 1992, 1993).

Although the Environmental Impact Statements for Outer Continental Shelf lease sales have in the past focused heavily on the probability of oil spills and the effect of spills on the physical environment, the public often focuses more heavily on the social impacts of Outer Continental Shelf activity (Freudenburg and Gramling 1994a). It is precisely the type of overadaptation that occurred in the Gulf that many of the opponents of the federal Outer Continental Shelf program want to avoid (Freudenburg and Gramling 1992a, 1994a; Gramling and Freudenburg 1992a; Wilson 1982), but as a result of past policy, Minerals Management Service has little information on this subject. Past research funded by Minerals Management Service has resulted in much valuable knowledge for the scientific community, but is of much less value to the agency as the program is currently administered. Because the requirements for data are not grounded in a realistic planning process, this allows the research program to drift, and as a consequence fail to target the relevant data (National Research Council 1989, 1992, 1993).

To pick only one example, a considerable percentage of Minerals Management Service's research budget has been spent on physical oceanography, both because of the expense of oceanographic research and because the acquisition of this data would appear to be reasonable. One would think that such information would be very valuable to the agency, since the physical transport of oil by currents is a critical consideration in the event of an oil spill. However, I would argue that the information is not useful under the current leasing policy nor, under the currently leasing policy, can it be used by Minerals Management Service, at least not for leasing decisions. With area-wide leasing, virtually the entire area is available in each sale, and currently there appears to be little that decides whether or not a tract is leased other than whether it meets the minimum bid. There is certainly no publicly visible connection between the data Minerals Management Service has collected on physical oceanography and the consideration of whether a tract should be leased. That is, since virtually all tracts that receive a sufficiently high bid are leased, the decision to lease a tract is not based on whether or not an oil spill from it would damage sensitive areas. In addition, there have been no realistic assessments of which areas are sensitive nor even any clear definitions of what variables would be used to determine sensitivity. If tracts are not to be excluded based on higher probability of major spills impacting certain areas, then the information on where the oil would go is not really relevant. This is not to say that the information would not be useful under a more comprehensive management program, such as the one proposed here, but simply that the current leasing structure precludes the input of such data into the decision-making process.

Although data on currents would assist tracking a spill and would theoretically be useful in spill response as a practical matter, at least with a major spill, knowing where the oil is going would probably make little difference. The *Exxon Valdez* disaster and experience from numerous other major marine spills has clearly demonstrated that there is little that can be done to contain or collect oil spilled on the open sea (Clarke 1990). Both of the previous two points brings up the third specific issue that Interior must deal with, the current leasing policy.

3. Area-wide leasing. Area-wide leasing appears to have resulted in two accomplishments, the transferring of large amounts of public property into private hands and for a limited period, making a considerable amount of money for Interior. What area-wide leasing specifically *has not* accomplished is increased production from the Outer Continental Shelf or reduced U.S. dependency on imports. The major oil companies like area-wide leasing, not only because of the cheap leases available, but also because their superior capital gives them an edge under this leasing strategy.

With the leasing structure in effect before area-wide leasing, industry nominated tracts to be included in a sale, Interior added tracts and then advertised in advance which tracts would be available for lease at the next sale. What this meant was that only a fraction of the total unleased area was offered at each sale and that the choices of which tracts to offer were made on the basis that either industry or Interior had reason to believe that there might be petroleum under them. Thus, individuals or corporations could bid on tracts, even if they had no independent data, with a fair chance that resources would be found later. In other words, this practice encouraged competition for the tracts offered for sale, a basic goal specified under the Outer Continental Shelf Lands Act. Under area-wide leasing the entire western and central Gulf are on the block at each sale. Thus, only those with independent data (i.e., data they have gathered) have real knowledge as to the potential for most tracts, and only the major oil companies can afford the type of extensive surveys necessary to examine even a portion of the entire central Gulf. As a result, the leasing process is biased toward the major oil companies and has placed the Department of Interior in the position of offering for sale a commodity when only the purchaser has a good idea of its potential value.

The result has been a marked decrease in competition. Between October of 1954, the first federal lease sale, and March of 1983, when area-wide leasing went into effect, the Department of Interior leased 3,559 tracts in the Gulf of Mexico (an average of about 122 a year) and received 12,921 bids on those tracts, for a ratio of 3.63 bids per tract leased. Beginning with the first area-wide lease sale in the Gulf in May of 1983 and continuing through the 1990 sales in the central and western Gulf, the Department of Interior leased 6,368 tracts in the Gulf (an average of about 796 a year) but received only 9,746 bids, for a ratio of 1.55 bids per tract leased (Gould et al. 1991: 75). As a result of this decrease in competition, those who have independent data, even when they are interested in a tract, especially in frontier areas, can submit smaller bids since the probability of another bid on that tract is relatively low. The offshore leasing structure in the Gulf is currently (1994) under review, and Minerals Management Service has called for public comments on its leasing policy in the Gulf (Krapf 1994). Whether or not the agency can adopt a policy that takes a long-term management approach, as opposed to a short-term transfer of ownership approach, is the most fundamental question under consideration.

4. The Intent of the Outer Continental Shelf Lands Act Amendments. More broadly than the impact in the Gulf, the direction of the research efforts aimed at assessing and monitoring the effects of the Outer Continental Shelf leasing program or the modification of area-wide leasing, the overall intent of the Outer Continental Shelf Lands Act Amendments must be addressed. It is quite clear that Congress intended, as a result of the amendments, for Outer Continental Shelf development to become more of a partnership between the Department of Interior and the affected coastal states and communities (Legislative History, P.L. 95-372, p. 54). The impetus for the current gridlock lies with the initial implementation of the amendments by an administration that simply did not agree with their intent (Freudenburg and Gramling 1994b). Watt's attempts to bowl over the resisting states and communities through the courts simply led the states to enter the political arena, dig in their heels, and, as resistance mounted through the 1980s, move the Outer Continental Shelf leasing decisions from the Department of Interior to the United States Congress. Any attempt at reviving the national Outer Continental Shelf leasing program must deal with the underlying concerns that have resulted in the actions to date.

5. Energy Independence. Perhaps the most difficult issue for Interior to deal with is the two-decade-old assumption that Outer Continental Shelf development can lead to some form of energy independence. I can envision *no* Outer Continental Shelf development scenario that can bring about energy independence for the United States or even significantly reduce its dependence on foreign imports, and I challenge anyone to come up with a credible scenario. Production of oil on the Outer Continental Shelf peaked over two decades ago at 418.5 million barrels in 1971, and by 1990, in spite of the fact that the entire central and western Gulf of Mexico (the area with the most potential) had been on the block annually, production was down to 324.4 million barrels (American Petroleum Institute 1993), 12 percent of total U.S. production. The United States has been the world's largest consumer of oil from the time of Edwin L. Drake's first oil well to the present. Our production of oil, however, is a different story. Total U.S. production of oil peaked at 3.5 billion barrels in 1970; by 1990 it had fallen to 2.7 billion barrels (American Petroleum Institute 1993). Today, U.S. crude oil reserves have been falling for over two decades and account for only 2.47 percent of world reserves. Two thirds (66.37%) of world crude oil reserves are in the Near East (American Petroleum Institute 1993). Our long history of consuming our own resources has left us with declining reserves and a growing dependence on imports that no amount of consuming our limited reserves faster will turn around.

LOOKING AT THE FUTURE

The Outer Continental Shelf is certainly not exhausted, and there are even positive signs for production on it. New three-dimensional seismic techniques allow better visualization of geological structure and have already resulted in the location of new reservoirs from areas in the Gulf of Mexico thought to be depleted or in decline. The recent announcement of two large discoveries, one in the Chukchi Sea and one in the central Gulf, are indicators that reserves on the Outer Continental Shelf may be larger than thought (but still not large enough to bring on energy independence). In addition, the 1992 Energy Policy and Conservation Act (42 U.S.C. § 6201) provides for greater use of alternative fuels for the federal vehicle fleet and significant tax incentives for service stations that install alternative fuel refilling facilities. The most likely alternative fuel for automobiles is natural gas, and these initiatives to encourage natural gas usage will eventually lead to more of these resources in the Gulf being tapped.

However, unless the federal Outer Continental Shelf program becomes a national one, its future, at least under current conditions, is very questionable. The decline in the Gulf of Mexico is inevitable, and with rising political pressures against it, there may well be fewer and fewer reasons for upper level policy makers to defend the current Outer Continental Shelf program. The revenues that the program takes in are declining steadily. Annual revenues from the Outer Continental Shelf peaked at almost $10 billion in the early 1980s but have since declined to about one third of that amount (Table E.3). Although primarily driven by the drastic decline in bonuses since area-wide leasing was put into effect, this trend, coupled with the slower but also steady drop in royalties, may signal an inevitable direction for the program. Given the current management of the Outer Continental Shelf and the current political climate, it is likely that the federal offshore program will continue to dwindle as a source of revenue for federal coffers and be increasingly costly politically, at least in the immediate future, for those outside the Gulf region who support it. If leasing the Outer Continental Shelf continues as the primary objective of Minerals Management Service, then the very existence of the agency may well be threatened. So, perhaps it is time to stop and carefully review how best to use this publicly owned finite resource in the future, in a way that considers the fortunes of the country, the regions of the country contiguous to the resources, and the existing infrastructure in the Gulf of Mexico, rather than just those of the major oil companies.

What then are possible new goals for such a massive program, if energy independence and revenue generation are no longer viable goals? There appear to be two possibilities. First, the Outer Continental Shelf represents a finite resource and, more importantly, the majority of the publicly owned oil and gas reserves of the country. Given this, a management program that looks at conservation and carefully thought out use of these public resources, as opposed to the most rapid exploitation possible, could be turned into a focus of Interior's stewardship of the Outer Continental Shelf. It would not be an easy task to change the direction of the program or to effectively manage these vast offshore lands.

For the former, it would require a major mind shift at the top levels of Interior and Minerals Management Service, but changes

Table E.3. Revenue Sources and Total Revenue from the Outer Continental Shelf (x 1,000) 1954–1990

YEAR	BONUSES	MINIMUM ROYALTIES	RENTALS	SHUT-IN GAS PAYMENTS	ROYALTIES	TOTAL
1953	0.0	0.0	1,359.6	30.7	967.9	2,358.2
1954	140,969.0	0.0	3,855.3	87.0	2,749.0	147,660.3
1955	108,528.7	0.0	3,406.4	122.0	5,140.0	117,197.1
1956	0.0	0.0	4,006.2	80.0	7,629.4	11,715.6
1957	0.0	68.6	3,270.1	110.3	11,391.2	14,840.2
1958	0.0	184.4	2,420.6	121.2	17,423.9	20,150.1
1959	89,747.0	171.0	2,285.7	85.0	26,540.0	118,828.7
1960	282,717.1	317.0	3,603.1	49.4	36,807.7	323,494.3
1961	0.0	314.1	3,073.9	37.1	46,733.7	50,158.8
1962	489,481.1	517.7	8,412.2	62.2	65,255.2	563,728.4
1963	12,807.3	668.3	8,435.2	53.0	75,373.9	97,337.7
1964	95,874.3	820.3	9,798.6	45.8	86,535.3	193,074.3
1965	33,740.3	1,072.7	8,731.4	38.5	99,656.3	143,239.2
1966	209,199.9	1,367.3	6,869.3	41.7	132,849.9	350,328.1
1967	510,109.7	1,891.5	6,208.9	41.4	153,432.4	671,683.9
1968	1,346,487.3	2,145.2	8,230.8	52.3	196,491.4	1,553,407.0
1969	111,660.7	1,923.6	8,312.6	41.7	235,681.8	357,620.4
1970	945,064.8	1,745.9	8,607.9	47.7	276,521.7	1,231,988.0
1971	96,304.5	1,891.0	7,742.0	32.3	340,634.9	446,604.7
1972	2,251,347.6	2,019.5	7,984.9	49.6	353,581.9	2,614,983.5
1973	3,082,462.6	2,391.2	8,948.8	52.7	389,735.8	3,483,591.1
1974	5,022,860.8	2,048.4	13,532.8	32.6	536,019.0	5,574,493.6
1975	1,088,133.2	2,085.9	17,522.0	39.5	594,725.5	1,702,506.1
1976	2,242,898.5	2,128.3	23,370.5	38.4	680,392.7	2,948,828.4
1977	1,568,564.7	1,679.0	19,830.0	21.1	890,469.8	2,480,564.6
1978	1,767,042.1	2,207.2	21,512.7	4.0	1,139,198.2	2,929,964.2
1979	5,078,861.7	2,088.5	20,287.3	6.6	1,512,017.7	6,613,261.8
1980	4,204,640.3	2,291.4	19,062.5	0.0	2,132,528.7	6,358,522.9
1981	6,652,980.9	2,250.3	21,731.0	0.0	3,287,279.4	9,964,241.6
1982	3,987,490.0	2,393.8	20,055.2	0.0	3,814,871.6	7,824,810.6
1983	5,749,016.4	4,463.7	32,463.1	0.0	3,454,318.1	9,240,261.3
1984	3,928,876.3	3,458.1	35,607.8	0.0	3,914,724.6	7,882,666.8
1985	1,557,650.7	4,067.3	61,999.1	0.0	3,612,782.8	5,236,499.9
1986	187,094.7	5,176.6	52,958.2	0.0	2,533,376.7	2,778,606.2
1987	497,247.0	21,399.3	74,642.7	0.0	2,337,034.9	2,930,323.9
1988	1,259,548.7	16,822.8	62,767.8	0.0	2,057,756.2	3,396,895.5
1989	645,646.9	38,890.9	79,247.7	0.0	2,118,785.5	2,882,571.0
1990	584,301.9	19,298.2	79,339.1	0.0	2,630,318.1	3,313,257.3
Totals	55,829,356.7	152,259.0	781,493.0	1,423.8	39,807,732.8	96,572,265.3

Source: American Petroleum Institute 1993.

associated with real management of the Outer Continental Shelf would be even more profound. To actually begin to manage the Outer Continental Shelf would probably require Interior, or more properly Minerals Management Service, to begin to utilize some of the revenues generated by the Outer Continental Shelf leasing program to directly contract for seismic exploration to more clearly identify the existing resources of the Outer Continental Shelf and, once these data have been processed, for exploratory drilling to delineate those resources. What a radical but hardly new idea that the agency entrusted with the resources on public lands should take an active role in their assessment and should know what those resources are in order to plan for their future use or conservation. Not only would such a new focus represent rational planning for the future, but it would mean that the enormous amount of scientific knowledge that the Minerals Management Service's Environmental Studies Program has generated would become relevant, useful information in that effort and, furthermore, that the agency would have clear criteria to evaluate the gaps in the data in order to direct future research! While the development of the Outer Continental Shelf has been argued as a matter of "national security," I would argue for its careful management and conservation for the same reason. The limited reserves of the Outer Continental Shelf may well be needed in the future for the transition from oil to alternative energy sources that ultimately must occur if the United States is to remain a superpower. A superpower dependent on oil but with inevitably dwindling domestic reserves is dangerously balanced as the OPEC 1973–1974 embargo demonstrated.

A second reasonable focus of an Outer Continental Shelf management program would be the preservation of the existing infrastructure in the Gulf of Mexico through a much more carefully targeted leasing program, integrated with a monitoring program like the one required under the Outer Continental Shelf Lands Act Amendments. The offshore infrastructure in the Gulf is the most developed in the world and as such represents an important national resource. As Figure 5.1 (chapter 5) shows, increasingly offshore oil is becoming the dominant source of new production worldwide. The huge continental shelves off China, Korea, and northern Russia are largely unexplored because those countries lack the technological expertise to explore them and cold war tensions have denied U.S. and European access. We can allow other nations to develop this potential, as some are already doing, or we can export technology and skills from the United States to support offshore development worldwide. The scheduling of offshore work makes it possible for skilled labor from the Gulf to work virtually anywhere in the world, while continuing to reside in communities in the United States (Gramling 1989), and the deeper waters of the Gulf of Mexico have the potential to provide a proving ground for future offshore technology.

While the potential is there, the offshore sector in the Gulf has been rocked by the boom associated with global events and an aggressive Federal leasing program and the inevitable bust that followed it. Many of the businesses that were started under the unrealistic economic environment of the boom years are gone. Those that remain are recovering from the 1986–1987 bust (Krapf 1994), and today the infrastructure appears to be more in line with the current level of development worldwide. The Gulf infrastructure does not need another uncontrolled boom nor does it need a drastic curtailment of activities if it is to remain viable and in a competitive position in an increasingly global market. As Freudenburg and Gramling (1994a) have noted, there are few hidden agendas with regard to positions toward offshore development in the United States. Development is supported as strongly in the central and western Gulf as it is opposed elsewhere. Given that support in the Gulf, an opportunity exists to maintain a viable national expertise through a carefully designed leasing and monitoring program, one that may become increasingly important from a world economic perspective.

While the first potential redirection noted would require some restructuring of Minerals Management Service, the second would not. Minerals Management Service already has a scientific committee that is used to advise on the funding of studies. A similar committee with Gulf of Mexico expertise could be used to focus on the question of how to best lease Gulf of Mexico resources in order to maintain the existing Gulf infrastructure. Such a committee could work with the existing regional technical working groups that Minerals Management Service has established in each region.

Are such goals doable and viable, and if so, within Interior and Minerals Management Service? I think so, but the first two questions are easier than the third. The legal framework is certainly there in the broad mandates of the OCSLAA, and the major oil companies have been managing their reserves for over a century. In fact one could argue that they have been managing their resources far better than the Department of Interior, to date, has been managing ours. One has only to look at Norway and Newfoundland and the research and management programs already underway there to realize that much more targeted national offshore management is possible. A U.S. program does not, and perhaps could not, emulate the structure of those approaches, but it could be established with some of the same management goals and objectives (Nelsen 1991).

Could Minerals Management Service do this? While large federal agencies have a lot of momentum and change does not come easily, there is evidence that Minerals Management Service is flexible enough that changes, such as those I am proposing, could occur. At least the Environmental Studies Program[3] in Minerals Management Service has demonstrated such flexibility, and it is the focus of that program that would have to change the most. Starting in late 1989, two separate reports from the National Academy of Science (National Research Council 1989, 1992) pointed out that the impact of Outer Continental Shelf activities on the human environment along the Gulf of Mexico, which I have argued has been the most significant impacts of the Outer Continental Shelf program, has received virtually no attention, assessment, or monitoring by Minerals Management Service.

Responding to this critique, Minerals Management Service sponsored a workshop in New Orleans in the fall of 1992. This was not an ordinary workshop. From the beginning, the workshop was planned in conjunction with scientists from Minerals Management Service. The workshop drew on the expertise of some of the best known social scientists in the country, indeed world, in the area of social and economic impacts of natural resource development and coastal, marine, and natural resource policy. In four days, utilizing a carefully orchestrated combination of plenary meetings and breakout groups and starting with a blank slate, these scientists in cooperation with those from Minerals Management Service designed an entire five-year social science research agenda for Minerals Management Service in the Gulf (Gramling and Laska 1993). Furthermore, most of those connected with the project and those who have later reviewed it agreed it was a success. Several of the projects are already underway as of this writing, and those who are knowledgeable in the agency give every indication that further projects will soon start. This does not indicate to me a rigidity that would be unable to refocus but rather a remarkably fast response time for a program of that size. This is particularly true given the nature of the annual budgetary cycle and the need to plan major projects well in advance.

Indeed my experience with Minerals Management Service has led me to believe that the problems that Minerals Management Service now faces are not because of the rank and file employees at the agency or even with midlevel administration. Most of the individuals I know at these levels are flexible people. Rather, I believe, Minerals Management Service is in its current position because of a succession of directors and, even beyond that, secretaries of Interior, who have dogmatically and unswervingly followed the failed policies started by Richard Nixon and compounded by James Watt. Watt has been gone for a decade; it is time to look beyond him. In short, I am proposing a larger, not smaller, role for the Minerals Management Service, but a very different one.

How would such a change be initiated? I believe the workshop in the Gulf (Gramling and Laska 1993) offers a model. It would require a series of local and regional workshops leading up to national ones. It would require local and regional representation and travel by those local and regional representatives to other areas, not necessarily in order to agree with the positions of those in other regions but in order to at least understand them. It would require careful planning and a willingness to listen and learn by all concerned. It would take time and often require painful reiteration and repetition. It would require careful targeting of areas to be leased, once a program was outlined, and careful managing of leases through exploration and development. Most of all, it would require longer-term thinking. But what are the alternatives? The current national program is obviously broken, and unless something is done to fix it, the future is not bright.

Over the past two decades in examining the federal offshore program, I have met a lot of people, working in coastal areas of the United States. Many of them appear to be unalterably opposed to offshore oil and gas activities in their areas and would see such a move by Interior or Minerals Management Service as a ploy. I must confess I have come to share many of these people's perspectives concerning their regions. That is, I don't think it currently makes sense to try to develop areas like Bristol Bay, Alaska, the largest salmon hatchery on the planet; the spectacular northern California coast; or the unique Florida Keys, and there are others. Any risk to these areas is, to me, unacceptable. But I am also aware that never is a very long time. Given the improbability that our political and economic leaders will look beyond their current two-, four-, or six-year term or their tenure as CEO, and consequently that the Unites States will, until literally forced to, seriously investigate alternative energy sources, and given the thirst for energy that we as Americans have, the grim political and economic realities will probably mean that inevitably the oil on the Outer Continental Shelf, all of it, will be developed. Nor, if the demand is great enough, will current or future Congressional action stop it. There is nothing that Congress can do that it cannot later undo. Thus, the question may be when and how, not if. It may be generations in the future, but far better that it be done with full knowledge of the potential there, with a technology that continues to improve and learn, and even with the input from and knowledge of people who totally oppose it than under the current situation.

Nor are oil and gas the only Outer Continental Shelf management issues. From manganese and cobalt in the Pacific, placer gold off Alaska, to the immense deposits of sand in federal waters immediately south of Louisiana's eroding barrier islands, there are many decisions facing whatever Outer Continental Shelf management structure that emerges. All of these are publicly owned resources, and failure to take that into account will lead to continued problems. Although he was speaking specifically to oil and gas considerations, there is a useful lesson in the cautionary note of a new Secretary of Interior Bruce Babbitt's first address at the Minerals Management Service headquarters office:

> . . . we're in this fix because we did a lot of leasing in the past without really providing the consensus—the political and public consensus to back up the leasing decision, it all unraveled, and now we're faced with going back with some really major economic consequences. The cautionary tale I think from here on forward is we've got to try to make certain that leasing decision are made through *a very public process.* . . . (emphasis in original, Cedar-Southworth 1993: 4)

Appendix

Selected Portions of the Outer Continental Shelf Lands Act Amendments

Panel 1: State and Local Participation

Purposes

Sec. 102. The purposes of this Act are to—

(1) establish policies and procedures for managing the oil and natural gas resources of the Outer Continental Shelf which are intended to result in expedited exploration and development of the Outer Continental Shelf in order to achieve national economic and energy policy goals, assure national security, reduce dependence on foreign sources, and maintain a favorable balance of payments in world trade;

(2) preserve, protect, and develop oil and natural gas resources in the Outer Continental Shelf in a manner which is consistent with the need (A) to make such resources available to meet the Nation's energy needs as rapidly as possible, (B) to balance orderly energy resource development with protection of the human, marine, and coastal environments, (C) to insure the public a

fair and equitable return on the resources of the Outer Continental Shelf, and (D) to preserve and maintain free enterprise competition; . . .

(4) provide States, and through States, local government which are impacted by Outer Continental Shelf oil and gas exploration, development, and production with comprehensive assistance in order to anticipate and plan for such impact, and thereby to assure adequate protection of the human environment;

(5) assure that States, and through States, local governments, have timely access to information regarding activities on the Outer Continental Shelf, and opportunity to review and comment on decisions relating to such activities, in order to anticipate, ameliorate, and plan for the impacts of such activities;

(6) assure that States, and through States, local governments, which are directly affected by exploration, development, and production of oil and natural gas are provided an opportunity to participate in policy and planning decisions relating to management of the resources of the Outer Continental Shelf;

(7) minimize or eliminate conflicts between the exploration, development, and production of oil and natural gas, and the recovery of other resources such as fish and shellfish; (43 USC 1802)

Panel 2: Assessment and Monitoring of Impacts

The Secretary shall conduct a study of any area or region included in any oil and gas lease sale in order to establish information needed for assessment and management of environmental impacts on the human, marine, and coastal areas which may be affected by oil and gas development in such area or region . . .

Subsequent to the leasing and developing of any area or regions, the Secretary shall conduct such additional studies to establish environmental information as he deems necessary and shall monitor the human, marine, and coastal environments of such area or region in a manner designed to provide time-series and data trend information which can be used for comparison with any previously collected data for the purpose of identifying any significant changes in the quality and productivity of such environments, for establishing trends in the areas studied and monitored, and for designing experiments to identify the causes of such changes. (43 USC 1346 Sec 20 (a)).

Notes

CHAPTER 1. DEVELOPING THE DEMAND

1. And might be noted was the reason for one in the Persian Gulf in 1991.

2. The hill proved to be the limestone cap over a salt dome and led to the search for other such domes, which were common along the Gulf Coast.

3. An image actively fostered by some of the early Texas leaders such as Joseph Cullinan, president of the Texas Company (Sampson 1975)

4. This was changed to 27.5 percent of gross income in 1926 to alleviate the necessity to adjudicate each claim (see Nash 1968: 85–86 for a discussion), and again in 1969 to 22 percent.

5. It was during this period of almost frantic efforts to develop foreign sources of oil (see Nash 1968: chapter 2) that Lake Maracaibo in Venezuela came into production. This was a major development point in offshore technology.

6. For a detailed explanation of how the process worked, see Ghanem (1986).

7. See Nash (1968: chapter 5) for a detailed discussion.

8. Ickes was one of the most powerful Secretaries of Interior in the history of the United States and the longest in office to date,

serving into the first year of the Truman administration. He was an ardent believer in conservation and the protection of public lands, and frequently came into conflict with the oil industry. See Solberg (1976) for a brief description, Ickes (1943, 1953) for his own account of this period.

9. The consortium members agreed not to act independently in an area covered by the "red line agreement" of 1928 (so named because during the negotiations one of the principles literally walked to the map and drew a red line around the area the agreement would cover). The agreement covered Turkey, Cyprus, Lebanon, Syria, Jordan, Iraq, Saudi Arabia, and the remainder of the Arabian Peninsula, except for Kuwait. Significantly, it did not cover Iran.

CHAPTER 2. THE EARLY TECHNOLOGY AND POLITICS OF OFFSHORE OIL

1. Later merged with Creole Petroleum Corporation, which in turn was bought out by Standard of New Jersey in order to gain access into Venezuela (Lankford 1971; Sampson 1975)

2. Lake Pelto, while initially an enclosed coastal bay, is now, due to coastal erosion, an arm of Terrebonne Bay protected from the open Gulf by a series of barrier Islands (Isles Dernieres) off the Louisiana coast.

3. Cypress logging as it was practiced in the coastal areas also frequently required digging canals for access.

4. For a detailed background see Bartley (1953), for a more concise discussion see Nash (1968).

5. The continental shelf is the comparatively shallow waters between the shore, and the continental slope which is the steeper drop into the deep ocean basin. The width of this shelf varies around the world from only several miles to several hundred miles.

6. Interestingly, one of them proposed that the states should share in the revenue from royalties derived off federal offshore

leases, at the same 37.5 percent that were in effect for federal on-shore lands within states (*World Oil* 1953a). Although ultimately the states got none of the revenue from federal offshore leasing, this revenue sharing from offshore federal lands remains an issue to the present.

7. The Supreme Court ruled that the federal government had paramount right to the nations seas.

8. Ironically, in most cases this has not been the case. With the exception of the central and western Gulf of Mexico, it has been the federal government that has pushed to lease offshore and the states that have resisted it.

9. This preoccupation with Soviet influence in the Middle East, which lasted into the Bush administration, came, at least partly, from a letter that emerged when the German Foreign Office documents were captured in 1945. Soviet Foreign Minister Vyacheslav Molotov wrote his counterpart in the German government, "The territorial ambitions of the Soviet Union lie south of the Soviet Territory in the direction of the Persian Gulf." (Solberg 1976: 174)

10. The U.S.C. classification system refers to the volume and location within volume of the U.S. Code. Thus (43 U.S.C. § 1301–1315) refers to volume 43 of the U.S. Code, parts 1301–1315. While the U.S.C. classification refers to where the law is published, the other way in which federal laws are referenced is by a "Public Law" citation which is assigned according to the Congress which passed them, and the order in which they passed. Thus, for example, (P.L. 92-103) would designate the 103rd law passed by the 92nd Congress.

11. In the first federal sale off Louisiana the state offered some of the same tracts for bid after the federal government had advertised them for lease (*World Oil* 1954a).

12. Ironically, the leases were sold in the vicinity of the Florida Keys, and three exploratory wells were subsequently drilled. In the 1980s drilling off south Florida was to become the source of a major conflict between Florida and the Department of Interior (see chapter 7).

CHAPTER 3. MOVING OFFSHORE

1. Jackets are built on their side and barged offshore also on their side. Once in position the specialized barge(s) is selectively flooded, tipping the jacket upright.

2. By contrast the Sears Tower in Chicago is 1,454 feet tall, and the World Trade Center and Empire State Building, 1,353 and 1,250 respectively.

3. The difference between planing and displacement hulls is both a function of performance and design. Planing hulls usually have sharp entries ("V" bows) to cut through waves and are relatively flat over the back two thirds of the bottom. The propeller shaft(s) extends through the bottom of the hull and can thus push the boat up to where it is planing over the surface of the water and still maintain a bite in the water. Planing hulls are fast but necessarily more top heavy than displacement hulls. Displacement hulls are more rounded with lower centers of gravity. The propeller shaft(s) extends through the transom of the vessel (like a ship) and thus cannot push the vessel up to plane over the surface. Displacement hulls are slower than planing hulls but are more seaworthy and can carry heavy loads.

4. Gravity-based platforms are constructed of giant reinforced hollow caissons, which are turned upright and fastened to one another to form a rectangular structure, similar in shape to a grain elevator. From the top of these caissons a comparatively slender column(s) supports a deck on which production machinery and crew quarters are located. When the caissons are empty they float upright and can be towed to the offshore production site. There some of the caissons are flooded and the structure sits on the sea floor. Some of the caissons may be used for the storage of crude oil. The structure is designed so that the caissons are below the surface and wave action and the deck is held well above the surface and wave action.

5. Some of the more recent leases in more difficult terrain (deeper water or in Alaska) are for ten years.

6. The survey was a stratified random sample of businesses in east St. Mary Parish, stratified by Standard Industrial Classifica-

tion Code (U.S. Department of Commerce 1979) and number of employees. All employees of each sample firm were asked to complete a questionnaire involving basic social and economic characteristics (e.g., age, sex, marital status, place of residence, income, occupation, number of miles traveled to work, etc.). Of 152 sampled firms 107, or 70.4 percent, cooperated with the survey, resulting in 1,560 completed employee questionnaires, or about 11.1 percent of the total estimated labor force in east St. Mary Parish. For details of the sampling, administration and coding procedures, see Gramling (1980a).

7. A situation where the individual is faced with two or more sets of norms.

8. The Santa Barbara spill provided at least partial impetus for the National Environmental Policy Act (NEPA), Earth Day, much of the subsequent environmental movement in the United States, and eventually for the Outer Continental Shelf Lands Act Amendments (OCSLAA) in 1978.

CHAPTER 4. POLITICAL STORM CLOUDS

1. The first production platform in over 100 feet of water was placed in the Gulf of Mexico four years earlier, in 1954 (Clark 1963).

2. Continental, Cities Service, and Phillips.

3. The first sale on what is now considered the Atlantic Outer Continental Shelf actually occurred in 1959, when twenty-three tracts were leased and three exploratory wells were drilled south of the Florida Keys without incident (see chapter 7). The wells were dry holes, and all have been plugged and abandoned.

CHAPTER 5. BOOM AND BUST IN THE GULF

1. At Bay Marchand block 2, Eugene Island block 126, Grand Isle blocks 16 and 47, Main Pass block 69, South Marsh Island block 73, South Pass blocks 24 and 27, Timbalier Bay (in State waters), and West Delta block 30.

2. More suitable for the rough North Sea waters.

3. "Hot shot" drivers deliver the more portable equipment used offshore (e.g., pumps, compressors, and specialized offshore drilling tools) between the source of the equipment and offshore staging areas. Most of the drivers have a pickup truck and a large trailer and work "on call" twenty-four hours a day. This economic niche is possible because of the necessity to get equipment offshore as quickly as possible in order not to shut down an expensive offshore operation.

CHAPTER 6. RISING POLITICAL CONTROVERSY

1. Even Louisiana objected to the prospect of area-wide leasing. At the direction of then governor Treen, Groat (1981) noted: "The State of Louisiana is concerned about the potential impacts of proposed changes in OCS leasing procedures that are designed to open the entire OCS area to exploration and development. Under the proposed leasing schedule it is likely that the industry would concentrate its initial efforts in areas of proven production such as the central Gulf. The resulting intensive development would create severe economic and environmental impacts in coastal Louisiana. This would also lead to a major increase in the rate of depletion of our most productive OCS area which is not in the best long term interest of the United States."

2. Interestingly the quote is from one of the "purposes" contained in the Outer Continental Shelf Lands Act Amendments (43 USC § 1802). Other "purposes" listed in this section are to balance the Outer Continental Shelf development with protection of the "human, marine and coastal environments" and to provide greater input into decision making for the states.

3. The report was never officially made public, but then Congresswoman Boxer obtained a copy and released it to the press in June of 1990.

4. Louisiana Coastal Resources Program (Louisiana's federally approved program under the Coastal Zone Management Act).

CHAPTER 7. THE FLORIDA CONFLICT

1. As noted earlier, Florida and Texas claimed and were subsequently awarded three marine leagues as the boundary of state waters in the Gulf of Mexico. All other state boundaries in all other U.S. waters lie at three miles offshore.

2. The 26th parallel, which crosses Florida just south of Naples, quickly became Florida's line in the sand.

3. When exploratory drilling was going forward off the Keys.

4. Culture is used in the sociological-anthropological sense of the word, referring to a set of shared norms, beliefs, and values.

5. See Gramling and Freudenburg (1996) for a more detailed comparison of development in southern Louisiana and Florida.

6. Attempts to lower the lake and develop farming and settlement in the region started in the 1920s. Following several disastrous floods during the 1920s and mid-1940s, the U.S. Army Corps of Engineers constructed a series of canals and water control structures in the late 1940s. While this has allowed increasing inroads into the Everglades, the growing population of south Florida has created a water shortage problem that the earlier construction does not alleviate.

7. One of the problems that emerged after Hurricane Andrew slammed into Homestead, Florida, was that because so many of the residents were first-generation Floridians, there was little in the way of the kinship support system that traditionally sustains individuals and families after a disaster.

8. 1,250,000,000 bls. x $20 ÷ 20 (years) ÷ 6 (royalty rate) x .125 (state/county cut) = $26,041,600 per year.

9. This sum would be "allocated to all coastal States within 200 miles of a given tract. [w]eighted inversely according to each State's minimum distance from the tract." "Within each State, 50 percent of the allocation goes to the State government, and the remaining 50 percent is distributed to eligible counties within 200 miles of the tract, weighted inversely according to each county's minimum distance from the tract." (Minerals Management Service

1990a:1). The governor would delegate "eligible" counties under this proposal, but they "must include all coastal counties and may include other counties within 60 miles of the coast" (Minerals Management Service 1990a: 1).

EPILOGUE

1. In spite of increased offshore leasing in the Gulf by 1994 the United States was importing slightly more than half of the oil consumed annually.

2. See Stallings et al. 1977; Gramling 1980b; Gramling 1983; Gramling and Brabant 1984, 1986a, 1986b; Gramling and Freudenburg 1990, 1992, 1994; Brabant and Gramling 1991; Freudenburg and Gramling 1993, 1994a)

3. The Environmental Studies Program was established in the Department of Interior, within the Bureau of Land Management, in 1973, at least partly to comply with the requirements of NEPA. Its mission was modified somewhat following the Outer Continental Shelf Lands Act Amendments in 1978. In essence the Environmental Studies Program funds research to support leasing and management decisions, and has spent over $500 million since its inception.

References

Albright, Charles F. and Philip L. McLaughlin. 1952 (December). Integrated drilling barge has maximum flexibility. *World Oil,* pp. 139–141.

Alderdice, Robert. 1969 (February). Boats and the offshore operator. *Offshore,* pp. 76–72.

American Petroleum Institute. 1993. *Basic Petroleum Data Book.* Washington DC: American Petroleum Institute.

Anderson, Jack and James Boyd. 1983. *Fiasco.* New York: Times Books.

Applebome, Peter. 1993 (19 April). Murders in Miami terrify foreign visitors. *New York Times,* E2.

Armstrong, Edward R. 1947 (27 January). Seadrome-type offshore foundation for continental shelf operations. *The Oil Weekly,* pp. 40–41.

Armstrong, Ted A. 1968 (March). Drillers fog to Santa Barbara. *Offshore,* 23–27.

Atkins, James E. 1973 (April). Oil: This time the wolf is here. *Foreign Affairs.*

Barnett, Harold C. 1989. The extent of social regulation: Hazardous waste cleanup and the Reagan ideology. *Policy Studies Review* 8:15–35.

Barrett, Barney. 1970. *Water Measurements of Coastal Louisiana.* New Orleans: Louisiana Wildlife and Fisheries Commission.

Bartley, Ernest R. 1953. *The Tidelands Oil Controversy*. Austin: University of Texas Press.

Beck Robert J. 1988. Forecast review. pp. 89–98 in *Oil and Gas Journal, Data Book*. Tulsa: Pennwell Publishing Company.

Beer, Peter. 1986. Keeping up with the Jones Act. *Tulane Law Review* 61:379–414.

——. 1991. Decision: *State of Louisiana et al.*, versus *Manuel Lujan*, United States District Court, Eastern District of Louisiana. Civil Action no. 91-2910.

Berger, Peter and Hansfried Kellner. 1964. Marriage and the construction of reality. *Diogenes* 46:1–23.

Berger, Peter and Thomas Luckmann. 1966. *The Social Construction of Reality*. Garden City: Doubleday.

Black, Murray D. 1966 (August). New subsea work chamber. *Offshore*, pp. 73–74.

Brabant, Sarah. 1984. Education in the Coastal Zone Parishes, pp. 135–164 in R. Gramling and S. Brabant (eds.) *The Role of Outer Continental Shelf Activities in the Growth and Modification of Louisiana's Coastal Zone*. Lafayette: U.S. Department of Commerce/Louisiana Department of Natural Resources.

——. 1993a. The impact of a boom/bust economy on poverty, pp. 161–194, in Shirley Laska, Vern Baxter, Ruth Seydlitz, Ralph Thyer, Sarah Brabant, and Craig Forsyth, *Impact of Offshore Oil Exploration and Production on the Social Institutions of Coastal Louisiana*. New Orleans: Minerals Management Service.

——. 1993b. From boom to bust: Community response to basic human need, pp. 195–208 in Shirley Laska, Vern Baxter, Ruth Seydlitz, Ralph Thyer, Sarah Brabant, and Craig Forsyth, *Impact of Offshore Oil Exploration and Production on the Social Institutions of Coastal Louisiana*. New Orleans: Minerals Management Service

Brabant, Sarah and Robert Gramling. 1985. Academic versus street definitions of poverty: implications for policy formulations. *Applied Sociology* 2:33–42.

——. 1991. The impact of a boom/bust economy on poverty. Presented at the annual meeting of the American Sociological Association, Cincinnati.

Brantly, J. E. 1971. *History of Oil Well Drilling*. Houston: Gulf Publishing Co.

Bunker, Stephen G. 1984 (March). Modes of extraction, unequal exchange, and the progressive underdevelopment of an extreme periphery: The Brazilian Amazon, 1600–1980. *American Journal of Sociology* 89:1017–64.

——. 1989 (December). Staples, links, and poles in the construction of regional development theories. *Sociological Forum* 4:589–610.

Caffrey, Evan T. 1990. Slicing the net: A legislative answer to the problem of seaman status under the Jones Act. *Tulane Maritime Law Journal* 14:361–380.

Carmichael, Jim. 1975. Industry has built over 800 platforms in the Gulf of Mexico. *Offshore* 36(5):83–90, 230–236.

Catton, William R. 1986. Homo colossus and the technological turn-around. *Sociological Spectrum* 6:121–148.

Catton, William R., Jr. and Riley E. Dunlap. 1980 (September-October). A new ecological paradigm for post-exuberant sociology. *American Behavioral Scientist* 24:15–47.

Cedar-Southworth, Donna. 1993. MMS welcomes new interior secretary Bruce Babbitt. *MMS Today* 2(3):1–4.

Centaur Associates. 1986. *Indicators of the Direct Economic Impacts due to Oil and Gas Development in the Gulf of Mexico*. New Orleans: Minerals Management Service.

Cicin-Sain, Biliana. 1986. Ocean resources and intergovernmental relations: An analysis of the patterns, pp 87–103 in M. Silva (ed.) *Ocean Resources and U.S. Intergovernmental Relations in the 1980s*.

Cicin-Sain, Biliana, Robert Gramling, Ralph Johnson, and Charles Wolf. 1992. The evolution of the federal OCS program: National and regional perspectives, pp. 107–137 in National Research Council, *Assessment of the U.S. Outer Continental Shelf Environmental Studies Program: III. Social and Economic Studies*. Washington: National Academy Press.

Cicin-Sain, Biliana and Robert Knecht. 1987. Federalism under stress: The case of offshore oil and California, pp. 149–176 in Harry Scheiber (ed.) *Perspectives on Federalism* Berkeley: Institute of Governmental Studies, University of California.

Clark, James A. 1963. *The Chronological History of the Petroleum and Natural Gas Industries.* Houston: Clark Book Co.

Clarke, Lee. 1990 (November). Oil-spill fantasies. *Atlantic* (November).

Cochran, T. G. and William Miller. 1942. *The Age of Enterprise.* New York: Macmillan.

Coleman, James S. 1957. *Community Conflict.* Glencoe, IL: Free Press.

Comeaux, Malcolm L. 1972. *Atchafalaya Swamp Life: Settlement and Folk Occupations.* Baton Rouge: Louisiana State University Press.

Craig, J. C. 1956 (August). Costly sea and air fleets service offshore rigs. *Drilling,* pp. 70–74.

Cram, Ira H. 1956 (June). Offshore development proving successful. *World Oil,* pp. 78–81.

Cronon, William. 1991. *Nature's Metropolis: Chicago and the Great West.* New York: W.W. Norton.

Darmstadter, J. and H. Landsberg. 1976. The economic background, pp. 15–37 in R. Vernon (ed) *The Oil Crisis.* New York: Norton.

Darrah, William C. 1972. *Pithole: The Vanished City.* Gettysburg, PA: William C. Darrah.

Davidson, Rollo E. 1955. Offshore drilling gets DeLong mobile platform. *Drilling,* p. 75.

Davis, Donald W. 1990. Living on the edge: Louisiana's marsh, estuary and barrier island population. *Louisiana Geological Survey* 40:147–160.

Davis, Donald W. and John L. Place. 1983. The Oil and Gas Industry of Coastal Louisiana and Its Effect on Land Use and Socioeconomic Patterns. Reston, VA: U.S. Department of Interior/ U.S. Geological Survey (Open File Report 83-118).

Drilling Staff. 1955a (December). Worlds deepest well. *Drilling,* p. 52.

Drilling Staff. 1955b (December). Offshore. *Drilling,* pp. 68–70.

Drilling Staff. 1955c (December). Derrick barge is engineered for heavy duty Gulf work. *Drilling,* p. 93.

Drilling Staff. 1956a (August). Move it over. *Drilling,* p. 83.

Drilling Staff. 1956b (August). Safety rules take effect on the outer continental shelf. *Drilling,* p. 83.

Drilling Staff. 1956c (August). Mud delivery offshore serve drilling rigs. *Drilling*, p. 83.

Drilling Staff. 1957 (August). Spring storms and hurricane threats. *Drilling*, p. 177.

Drilling Staff. 1958a (March). Marine drilling. *Drilling*, pp. 83–87.

Drilling Staff. 1958b (March). Drilling log. *Drilling*, p. 16.

Dunlap, Riley E. and Angela G. Mertig. 1992. *American Environmentalism: The U.S. Environmental Movement, 1970–1990*. Philadelphia: Taylor and Francis.

Durio, C. and K. Dupuis. 1980. Public utilities, pp. 258–277 in R. Gramling (ed.) East St. Mary Parish, Economic Growth and Stabilization Strategies. Baton Rouge: Louisiana Department of Natural Resources.

Durkheim, Emile. 1951. *Suicide*, translated by J. A. Spaulding and George Simpson. New York: Free Press.

Edmondson, Cameron. 1958 (February). Drilling log. *Drilling*, p. 19.

Engler, Robert. 1961. *The Politics of Oil: A Study of Private Power and Democratic Directions*. New York: Macmillan.

———. 1977. *The Brotherhood of Oil: Energy Policy and the Public Interest*. Chicago: University of Chicago Press.

Farley, R. C. and J. S. Leonard. 1950 (March). Hurricane damage to drilling platform. *World Oil*, pp. 85–92.

Farrow, R. Scott. 1990. *Managing the Outer Continental Shelf Lands: Oceans of Controversy*. New York. Taylor and Francis.

Feagin, Joe R. 1985. The global context of metropolitan growth. *American Journal of Sociology* 90:1204–1230.

Fedarko, Kevin. 1993 (20 September). Miami's tourist trap. *Time*, p. 71.

Feder, Judy. 1974 (December). All offshore areas of the world show gains and development. *Offshore*, pp. 51–53.

Federal Trade Commission. 1952. The International Petroleum Cartel. Washington, DC: U.S. Government Printing Office.

Fitzgerald, Edward A. 1987. *California V. Watt*: Congressional intent bows to judicial restraint. *Harvard Environmental Law Review* 11:147–201.

Flynn, John T. 1932. *God's Gold, the Story of Rockefeller and His Times*. New York: Harcourt Brace Jovanovich.

Forbes, Gerald. 1946. A history of the Caddo oil and gas field. *Louisiana Historical Quarterly* 29:3–16.

Forsyth, Craig. 1989. *The American Merchant Seaman and His Industry: Struggle and Stigma*. New York: Taylor and Francis.

Forsyth, Craig, W. B. Bankston, and J. H. Jones. 1984. Organizational and extra-organizational determinants of occupationally induced social marginality: A study of merchant seamen. *Sociological Focus* 17:325–336.

Forsyth, Craig and DeAnn Gauthier. 1991. Families of offshore oil workers: Adaptations to cyclical father absence/presence. *Sociological Spectrum* 11:177–202

———. 1993. Familial structural responses to cyclical father absence: The offshore oil family, pp. 209–246, in Shirley Laska et al. *Impact of Offshore Oil Exploration and Production on the Social Institutions of Coastal Louisiana*. New Orleans: Minerals Management Service

Forsyth, Craig and Robert Gramling. 1987. Feast or famine: Alternative management techniques among periodic father absence single career families. *International Journal of Sociology of the Family* 17:183–195.

Freudenburg, William R., Scott Frickel, and Robert Gramling. 1995. "Beyond the Society/Nature divide: Learning to think about a mountain." Sociological Forum (in press).

Freudenburg, William R. and Robert Gramling. 1992a. Community impacts of technological change: Toward a longitudinal perspective. *Social Forces* 70:937–957.

———. 1992b. Linked to what? A critical examination of economic linkages in an extractive economy. Presented at the Annual meeting of the Rural Sociological Society, State University, PA, August.

———. 1993. Socio-environmental factors and development Policy: Understanding opposition and support for offshore oil. *Sociological Forum* 8:341–364.

———. 1994a. *Oil in Troubled Waters: Perceptions, Politics, and the Battle Over Offshore Drilling*. Albany: SUNY Press.

———. 1994b. Bureaucratic slippage and the failures of agency vigilance: The case of the environmental studies program. *Social Problems* 41:501–526.

Freudenburg, William R. and Susan K. Pastor. 1992 (August). Public responses to technological risks: Toward a sociological perspective. *Sociological Quarterly* 33 (3):389–412.

Fuchs, Richard P., Gary Cake, and Guy Wright. 1981. *The Steel Island*. St. John's, Newfoundland: Department of Rural, Agricultural and Northern Development, Government of Newfoundland and Labrador.

Gagliano, S. M. 1973. *Canals, Dredging, and Land Reclamation in the Louisiana Coastal Zone*. Report 14: Baton Rouge: Center for Wetland Resources, Louisiana State University.

Ghanem, Skukri. 1986. *OPEC: The Rise and Fall of an Exclusive Club*. London: Routledge and Kegan Paul.

Gibbon, Anthony. 1954 (December). Here's the newest ship in the petroleum navy. *World Oil*, p. 140.

Giddens, Paul H. 1975. *Edwin L. Drake and the Birth of the Petroleum Industry*. Historic Pennsylvania Leaflet No. 21: Pennsylvania Historic and Museum Commission.

Gould, Gregory J., Robert M. Karpas, and Douglas L. Siltor 1991. *OCS National Compendium*. Herndon, VA: Minerals Management Service.

Governor's Office of Planning and Budget. 1987. *Draft Report: Scientific Review of Environmental Studies Conducted by the U.S. Department of Interior in Consideration of Oil and Gas Drilling Off Southwest Florida*. Tallahassee: Governor's Office of Planning and Budget.

Gramling, Robert. 1980a. Labor survey of East St. Mary Parish, pp. 80–108 in Robert Gramling (ed.) East St. Mary Parish, Economic Growth and Stabilization Strategies. Baton Rouge: Louisiana Department of Natural Resources.

Gramling, Robert (ed). 1980b. East St. Mary Parish, Economic Growth and Stabilization Strategies. Baton Rouge: Louisiana Department of Natural Resources.

Gramling, Robert. 1983. A social history of Lafayette Parish, pp. 8–52 in David Manuel (ed.) *Energy and Economic Growth in Lafayette, LA: 1965-1980*. Lafayette: The University of Southwestern Louisiana.

———. 1984. East St. Mary Parish: A Case Study, pp. 171–198 in R. Gramling and S. Brabant (eds.) *The Role of Outer Continental*

Shelf Activities in the Growth and Modification of Louisiana's Coastal Zone. Lafayette: U.S. Department of Commerce/Louisiana Department of Natural Resources.

———. 1989. Concentrated work scheduling: Enabling and constraining aspects. *Sociological Perspectives* 32:47–64.

———. 1992. Employment data and social impact assessment. *Evaluation and Program Planning* 15:1–7.

Gramling, Robert and Sarah Brabant (eds.). 1984. *The Role of Outer Continental Shelf Activities in the Growth and Modification of Louisiana's Coastal Zone*. Lafayette: U.S. Department of Commerce/Louisiana Department of Natural Resources.

Gramling, Robert and Sarah Brabant. 1986a. Boom towns and offshore energy impact assessment: The development of a comprehensive mModel. *Sociological Perspectives* 29:177–201.

———. 1986b. Reply to Richard Gale. *Sociological Perspectives* 29:511–515.

Gramling, Robert and Craig Forsyth. 1987. Work scheduling and family interaction: A theoretical perspective. *Journal of Family Issues* 8:163–175.

Gramling, Robert and William R. Freudenburg. 1990. A closer look at "local control": Communities, commodities, and the collapse of the coast. *Rural Sociology* 55(4):541–558.

———. 1992a. Opportunity-threat, development, and adaptation: Toward a comprehensive framework for social impact assessment. *Rural Sociology* 57:216–234.

———. 1992b. The *Exxon Valdez* oil spill in the context of U.S. petroleum energy politics. *Industrial Crisis Quarterly* 6(3):1–23.

———. 1996. "Crude, Coppertone and the Coast: Developmental channelization and the constraint of alternative development opportunities." *Society and Natural Resources* (forthcoming).

Gramling, Robert and Edward Joubert. 1977. The impact of outer continental shelf petroleum activity on social and cultural characteristics of Morgan City, Louisiana, pp. 106–143 in E. F. Stallings (ed.) *Outer Continental Shelf Impact, Morgan City, Louisiana*. Baton Rouge: Louisiana Department of Transportation and Development.

Gramling, Robert and Shirley Laska. 1993. *A Social Science Research Agenda for the Minerals Management Service in the Gulf of*

Mexico. New Orleans: U.S. Department of Interior/Minerals Management Service.

Gramling, Robert and T. F. Reilly. 1980. Education in East St. Mary Parish, pp. 109–116 in R. Gramling (ed.) *East St. Mary Parish, Economic Growth and Stabilization Strategies.* Baton Rouge: Louisiana Department of Natural Resources.

Gregory, J. N. 1955 (December). Sea legs. *Drilling,* pp. 74–75.

Groat, C. G. 1981 (15 September). Letter from Charles G. Groat, director of the Louisiana Geological Survey, to John L. Rankin, U.S. Department of Interior, Bureau of Land Management.

Hance, Billie Jo, Caron Chess, and Peter M. Sandman. 1988. *Improving Dialogue with Communities: A Risk Communication Manual for Government.* New Brunswick, NJ: Rutgers University Environmental Communication Research Program.

Hanna, Fares K. 1955 (August). Off-shore trends. *Drilling,* pp. 78–100.

———. 1957 (August). Marine Designs. *Drilling,* pp. 71–75.

Hansen, Thorvald B., Odd J. Lange, Håkon Lavik, and Willy H. Olsen. 1982. *Offshore Adventure.* Stavenger, Norway: Universitetsforlaget.

Hill, Jack O. 1966 (June). Marine pipeline pre-lay surveys and right-of-way investigation. *Offshore,* pp. 69–72.

House, J. D. 1984. Oil and the North Atlantic periphery: The Scottish experience and the prospects for Newfoundland. Presented at Conference for Oil and Gas Development in Have-Not Regions: Some Lessons for Nova Scotia. Gorgebrook Research Institute, St. Mary's University, Halifax, Nova Scotia. July.

Howe, Richard J. 1966a (March). Evolution of offshore mobile drilling units: Part I. *Offshore,* pp. 68–92.

———. 1966b (April). Evolution of offshore mobile drilling units: Part II. *Offshore,* pp. 76–92.

———. 1966c (May). Evolution of offshore mobile drilling units: Part III. *Offshore,* pp. 102–120.

Humble Oil, Production Department. 1946 (December). Big diesel electric drilling rig. *The Oil Weekly,* 30:28–31.

Ickes, Harold L. 1943. *Fighting Oil.* New York: Knopf.

————. 1953. *Secret Diary of Harold Ickes*. New York: Simon and Schuster.

Johnson, D. 1977. Municipal services, pp. 160–208 in E. F. Stallings (ed.) *Outer Continental Shelf Impacts*, Morgan City, Louisiana. Baton Rouge: Louisiana Department of Transportation and Development.

Johnson, Paul G. and Debora L. Tucker. 1987. *The Federal Outer Continental Shelf Oil and Gas Leasing Program: A Florida Perspective*. Tallahassee: Governor's Office Of Planning And Budget

Jones, Russell O., Walter J. Mead, and Philip E. Sorenson. 1979. The Outer Continental Shelf Lands Act Amendments of 1978. *Natural Resources Journal* 19:885–908.

Kaplan, Elizabeth R.. 1982. California: threatening the golden shore. pp. 3–28 in Joan Goldstein (ed.) *The Politics of Offshore Oil*. New York: Praeger.

Kaplan, Max. 1960. *Leisure in America: A Social Inquiry*. New York: Wiley.

————. 1975. *Leisure: Theory and Policy*. New York: Wiley.

Kastrop. J. E. 1950 (June). Dual rig platform reduces offshore drilling costs. *World Oil*, p. 149.

Kaufman, Burton I. 1978. *The Oil Cartel Case: A Documentary Study of Antitrust Activity in the Cold War Era*. Westport, CT: Greenwood Press.

Killian, Molly S. and Charles M. Tolbert. 1993. Mapping social and economic space: The delineation of local labor markets in the United States, pp. 69–79 in Joachim Singelmann and Forrest A. Deseran (eds.). *Inequalities in Labor Market Areas*. Westview Press: Boulder.

Kinzer, Stephen. 1993 (10 September). Germans wary on slaying of another Florida tourist. *New York Times* p. A16.

Kniffen, Fred B. 1968. *Louisiana: Its Land and People*. Baton Rouge: Louisiana State University Press.

Krapf, David. 1994 (March-April). On the rebound. *Work Boat* p. 56.

Kruger, Robert B. and Louis H. Singer. 1979. An analysis of the Outer Continental Shelf Lands Act Amendments of 1978. *Natural Resources Journal* 19:909–927.

Lacy, Ray. 1957. A fresh approach to offshore development. *Drilling*, pp. 119–120.

Lacy, Ray and John Estes. 1960 (November). How new underwater oil storage unit works. *World Oil*, pp. 92–93.

Lahm, Jack. 1966 (August). New methods, equipment progress in deep diving. *Offshore*, pp. 75–76.

Lankford Raymond L. 1971. Marine drilling, pp. 1358–1444 in J. E. Brantly (ed.) *History of Oil Well Drilling*. Houston: Gulf Publishing Co.

Laska, Shirley, Vern Baxter, Ruth Seydlitz, Ralph Thyer, Sarah Brabant, and Craig Forsyth. 1993. *Impact of Offshore Oil Exploration and Production on the Social Institutions of Coastal Louisiana*. New Orleans: Minerals Management Service

Lietz, Richard T. 1956 (August). Equipment leasing helps meet offshore rig needs. *Drilling*, pp. 76–77.

Llewellyn, Lynn G. 1981. The social costs of urban transportation, pp. 169–201 in I. Altman, J. Wohlwill, and P. Everette (eds.) *Transportation and Behavior*. New York: Plenum.

Lloyd, Jacqueline M. and Joan M. Ragland. 1991. *Information Circular No 107: Part II: Petroleum Exploration and Development Policy in Florida: Response to Concern for Sensitive Environments*. Tallahassee: Florida Geological Survey.

Logan, L. J. 1953 (June). Offshore development to be speeded up. *World Oil*, pp. 53–54.

Logan, R. J. and Cecil Smith. 1948 (July). Continental Shelf activity intensified. *World Oil*, pp. 37–39.

Londenberg, Ron. 1973 (September). Florida hearing draws opposition but outlook is good. *Offshore*, pp. 36–37.

Louisiana Department of Natural Resources. 1991 (29 July). *Information Supplement Lease Sale # 135*.

Lovejoy, Stephen B. and Richard S. Krannich. 1982 (Fall). Rural industrial development and domestic dependency relations: Toward an integrated perspective. *Rural Sociology* 47:475–495.

Lurie, Alison. 1981. *The Language of Clothes*. New York: Vintage.

Magnuson, W. G. and E. F. Hollings. 1975. *An Analysis of the Department of the Interior's Proposed Acceleration of Development of Oil and Gas on the Outer Continental Shelf*. U.S. Government Printing Office, Washington, D.C.

Maloney, David M. 1990. *Final Brief of the State of Florida: In the Matter of the Appeal of Unocal Exploration and Producing North America Inc. to the Consistency Objection of the State of Florida to the Proposed Plan of Exploration for Lease OCS-G 6491/6492 (Pulley Ridge Blocks 629 and 630.* United States Department of Commerce: Before the Secretary.

Manners, Ian R. 1982. *North Sea Oil and Environmental Planning: The United Kingdom Experience.* Austin: University of Texas Press.

Manuel, David P. 1980. East St. Mary Parish in the 1970s: The economics of a sustained energy Impact, pp. 44–58 in R. Gramling (ed.) *East St. Mary Parish, Economic Growth and Stabilization Strategies.* Baton Rouge: Louisiana Department of Natural Resources.

——. 1984. Trends in Louisiana OCS activities, pp. 27–40 in R. Gramling and S. Brabant (eds.). *The Role of Outer Continental Shelf Activities in the Growth and Modification of Louisiana's Coastal Zone.* Lafayette: U.S. Department of Commerce/Louisiana Department of Natural Resources.

Martin, Charles K. 1956 (January). That submarine drilling rig plan intrigues the industry. *Drilling,* pp. 87–88.

McGee, Ed and Carl Hoot. 1963. Mighty dredges, little known work horses of coastal drilling, producing, pipelining, now 25 years old. *Oil and Gas Journal* 61(9):150–59.

McLaughlin, Philip L. 1953 (April). Luxury and efficiency afloat. *World Oil,* pp. 164–165.

Mead, W., A. Moseidjord, D. Mauraoka, and P. Sorensen. 1985. *Offshore Lands: Oil and Gas Leasing and Conservation on the Outer Continental Shelf.* Pacific Institute for Public Policy Research.

Minerals Management Service. 1984. *Final Environmental Impact Statement, Proposed Oil and Gas Lease Sales 94, 98 and 102: Gulf of Mexico OCS Region.* New Orleans: Minerals Management Service

——. 1989. Visual no. 1: Historic leasing structure and infrastructure. [map] New Orleans: Minerals Management Service.

——. 1990a. *Fact Sheet on Impact Assistance Proposal.* no city: Minerals Management Service.

——. 1990b. *Federal Offshore Statistics.* Herndon, VA: Minerals Management Service

——. 1993. Data file provided by Minerals Management Service to the author in the spring of 1993, containing information on all leases in the Gulf of Mexico. Statistical manipulations were done by the author in order to obtain the results reported.

MMS Today Staff. 1992. $100 billion in revenues from offshore. *MMS Today* 2(1):6.

MMS Today Staff. 1993. MMS Director Sewell announces oil and gas incentives package. *MMS Today* 2(2):1–7.

Molotch, Harvey. 1970 (Winter). Oil in Santa Barbara and power in America. *Sociological Inquiry* 40:131–144.

Monroe, Dorothy (May). 1955. Magnolia adds offshore field. *Drilling*, 127.

Morgan, Gareth. 1986. *Images of Organizations*. Newbury Park, CA: Sage.

Morris, Allen. 1991. *The Florida Handbook*. Tallahassee: Peninsular Publishing.

Nash, Gerald D. 1968. *United States Oil Policy 1890–1964*. Westport CT: Greenwood Press.

National Research Council. 1989. *The Adequacy of Environmental Information for Outer Continental Shelf Oil and Gas Decisions: Florida and California*. Washington, DC: National Academy Press, National Academy of Sciences.

———. 1992. *Assessment of the U.S. Outer Continental Shelf Environmental Studies Program: III Social and Economic Studies*. Washington, DC: National Academy Press, National Academy of Sciences.

———. 1993. *Assessment of the U.S. Outer Continental Shelf Environmental Studies Program: IV. Lessons and Opportunities*. Washington, DC: National Academy Press, National Academy of Sciences.

Nelsen, Brent F. 1991. *The State Offshore: Petroleum, Politics, and State Intervention on the British and Norwegian Continental Shelves*. New York: Praegar.

Nicholson, Gordon B. 1941a (6 October). Spindletop discovery marked birth of modern oil industry. *The Oil Weekly*.

———. 1941b. Houseboat provides many comforts at marine location. *The Oil Weekly*: September 15 p. 30.

———. 1942 (10 August). Road construction in swamps presents difficulty. *The Oil Weekly*, pp. 25–32.

Noreng, O. 1980. *The Oil Industry and Government Strategy in the North Sea*. London: Croom Helm.

Norgress, R. E. 1947. The history of the cypress lumber industry in Louisiana. *Louisiana Historical Quarterly* 30:3–83.

Ocean Industry Staff. 1974. Marine transportation fleet. *Ocean Industry* 6:44–51.

———. 1976. Marine transportation fleet. *Ocean Industry* 11:49–66.

Offshore Staff. 1966a (August). Shell drilling in record water depth off Louisiana. *Offshore* p. 46.

———. 1966b (January). Boom forecast as: oil demand grows world-wide. *Offshore* p. 52.

———. 1966c (July). Big hike in energy demand. *Offshore* p. 74.

———. 1966d (January). Operators ante up 84 million more in U.K. North Sea quest. *Offshore* pp. 18–22.)

———. 1966e (February). Big floater outbound for Persian Gulf. *Offshore* p. 35.

———. 1966f (February). *Glomar North Sea* to drill off Iran. *Offshore* p. 37.

———. 1966g (February). *Sea Quest* outfitting for North Sea. *Offshore* p. 38.

———. 1966h (February). U.S. built rig to drill off Norway. *Offshore* p. 40.

———. 1966i (September). New sonar in subsea work. *Offshore* p. 68–70.

———. 1967a (February). Huge bid off Santa Barbara. *Offshore* p. 98.

———. 1967b (December). Long wait on California sale idles men, ships, rigs. *Offshore* p. 24.

———. 1967c (July). Giant jackup to drill off Denmark. *Offshore* p. 29–30.

———. 1967d (June). North Sea gamble looks better. *Offshore* p. 92–100.

———. 1967e (March). Phillips gets record breaking strike. *Offshore* p. 25.

———. 1967f (April). First BP gas moves ashore. *Offshore* p. 30.

———. 1967g (June). Gulf of Mexico is still booming. *Offshore* pp. 60–69.

———. 1967h (July). Offshore Louisiana 1/2 billion sale. *Offshore* p. 21–26.

———. 1967i (January). 4 platforms to work for shell. *Offshore* p. 38.

———. 1967j (January). Drilling platform construction boom. *Offshore* pp. 20–24.

———. 1967k (September). "Super" drillers—a new trend? *Offshore* pp. 25–26.

———. 1968 (January). Zapata off-shore courts foreign rig assignments. *Offshore* pp. 55–56.

———. 1969a (June). Gulf of Mexico logs exciting year. *Offshore* pp. 41–44.

———. 1969b (January). Deep sea drilling project completes second leg. *Offshore* pp. 67–72.

———. 1974 (November). Offering more leases is only part of the solution. *Offshore* p. 53.

Oil Weekly Staff. 1946a (February). World crude oil production, by countries, by year. *The Oil Weekly* 11:100–102.

———. 1946b (September). U-shaped drilling barge. *The Oil Weekly* 2:40–41.

———. 1946c (August). Modern floating hotel for marine rig crews. *The Oil Weekly* 26:44–45.

———. 1946d (October). Magnolia Company's open gulf test is below 8400 feet. *The Oil Weekly* 28:31–31.

Peters, T. J. 1978. Symbols, patterns and settings. *Organizational Dynamics* 7:3–22.

Reilly, T. F. 1980. Recreation, pp. 288–302 in R. Gramling (ed.) *East St. Mary Parish Economic Growth and Stabilization Strategies.* Baton Rouge: Louisiana Department of Natural Resources.

Riposa, Gerry. 1989. After Reagan's deregulation: State-federal conflict over energy policy. *Policy Studies Review* 8:36–54.

Russell, Michele L. 1990a. *Brief of the State of Florida: In the Matter of the Appeal of Union Oil Company of California to the Consistency Objection of the State of Florida to the Proposed Plan of Exploration for Lease OCS-G 6491/6492 (Pulley Ridge Blocks 629 and 630.* United States Department of Commerce: Before the Secretary.

———. 1990b. *Brief of the State of Florida: In the Matter of the Appeal of Mobil Exploration and Producing Inc. to the Consistency Objection of the State of Florida to the Proposed Plan of Exploration for Lease OCS-G 6520 (Pulley Ridge Block 799.* United States Department of Commerce: Before the Secretary.

Russell, R. J. 1942. Flotant. *Geographic Review* 32:74–98.

Sampson, Anthony. 1975. *The Seven Sisters: The Great Oil Companies and the World They Made.* New York: Viking.

Schempf, F. J. 1968a (December). Louisiana lease nets $94 million. *Offshore,* pp. 45–46.

————. 1968b (July). Rig building tops previous records. *Offshore,* pp. 23–26.

Seale, A. F. T. 1948 (May). Discovering oil 12 miles offshore in the gulf. *World Oil* pp. 113–116.

Senate Committee on Energy and Natural Resources. 1987 (9 February). Hearings, Domestic Petroleum Industry Outlook. Lafayette, LA.

Senate Subcommittee on Multinational Corporations. 1974. *Multinational Corporations and United States Foreign Policy.* Committee on Foreign Relations, Hearings, 93rd Congress.

Seydlitz, Ruth, Shirley Laska, Daphne Spain, Elizabeth W Triche, and Karen L. Bishop. 1993a. The impacts of the offshore oil industry on suicide and homicide rates. *Rural Sociology.* 58:93–110.

————. 1993b. Development, human capital, and economic health: An empirical examination, pp. 105–125 in Shirley Laska (ed.), *Impact of Offshore Oil Exploration and Production on the Social Institutions of Coastal Louisiana.* OCS Study MMS93-0007. U.S. Dept. of the Interior, Minerals Management Service, Gulf of Mexico OCS Regional Office. New Orleans, La.

Shrewsbury, R. D. 1945 (10 September). Deep sea drilling. *The Oil Weekly,* 33–45.

Shrimpton, Mark and Keith Storey. 1987. Fly-in: The benefits and costs of a new approach to developing remote mineral resources. Department of Geography, Memorial University of Newfoundland, St. John's, Newfoundland: Unpublished paper.

Sitterson, Joseph C. 1973. *Sugar Country: The Cane Sugar Industry in the South, 1753–1950.* Westport, CT: Greenwood Press.

Smircish, E. 1983. Concepts of culture and organization analysis. *Administrative Science Quarterly* 28:339–58.

Smith, Gregory C. 1992. *Final Brief of the State of Florida: In the Matter of the Appeal of Chevron, U.S.A. Inc. to the Consistency Objection of the State of Florida to the Proposed Plan of Exploration for*

Lease OCS-G 8336 (Destin Dome Block 97. United States Department of Commerce: Before the Secretary.

Smith, Robert. 1993 (September-October). Down and out: Platform removal is a promising area for Gulf OSV operators. *Work Boat,* p. 23.

Snell, Bradford C. 1974. American Ground Transportation. a report to Subcommittee on Antitrust and Monopoly of The Committee on the Judiciary, *The industrial Reorganization Act,* U.S. Senate, 93rd Congress, 2nd Session.

Solberg, Carl. 1976. *Oil Power.* New York: Mason.

Stallings, E. F., T. F. Reilly, R. B. Gramling and D. P. Manuel (eds.). 1977. Outer Continental Shelf Impacts: Morgan City, Louisiana. Washington, DC: Government Printing Office.

Stallings, E. F. and T. F. Reilly. 1980. Transportation—East St. Mary Parish, pp. 278–297 in R. Gramling (ed.). East St. Mary Parish Economic Growth and Stabilization Strategies. Baton Rouge: Louisiana Department of Natural Resources.

Steele, Valerie. 1985. *Fashion and Eroticism: Ideals of Feminine Beauty From the Victorian Era to the Jazz Age.* New York: Oxford University Press.

Steinhart, Carol E. and John S. Steinhart. 1972. *Blowout: A Case Study of the Santa Barbara Oil Spill.* Belmont, CA: Duxbury Press.

Sterrett, Elton. 1941 (20 October). 25-Mile under-water span in seafaring pipe line. *The Oil Weekly,* pp. 33–38.

Sterrett, Elton. 1948a. Giant rig for deepest marine drilling. *World Oil,* pp. 75–89.

———. 1948b (June). Amphibian drilling goes deep sea. *World Oil* pp. 77–82.

Storey, Keith, J. Lewis, Mark Shrimpton, and D. Clark. 1986. *Family Life Adaptations to Offshore Oil and Gas Employment.* Ottawa, Canada: Environmental Studies Revolving Funds Report.

Storey, Keith and Mark Shrimpton. 1986. A review of the nature and significance of the use of long distance commuting by Canadian resource industries. Research paper prepared for Review of Demography and the Implications for Economic and Social Policy, Health and Welfare Canada.

Taylor, Donald M. 1951 (November). Only four months required to lay biggest offshore line. *World Oil* pp. 223–224.

Tebeau, Charlton W. 1971. *A History of Florida*. Coral Gables: University of Miami Press.

The Times of Acadiana. 1993 (15 September). Petroleum Club of Lafayette celebrates 40th anniversary, p. 2.

Thobe, Susan. 1973 (August). Norwegians make big play for rig market. *Offshore*, 27–31.

———. 1974 (January). Oil explorers capture Destin Dome in spirited bidding. *Offshore*, pp. 39–41.

Thomas, Henry F. 1946 (4 February). A report on an expedition to the far north to locate rumored seeps and secure samples. *The Oil Weekly*, pp. 39–48.

Trow, Raymond B. 1952 (August). The wettest inch. *World Oil*, pp. 249–252.

Tubb, Maretta. 1977. 1977–1978 Directory of marine drilling rigs. *Ocean Industry* 12 (9):39–180.

———. 1978. 1978–1979 Survey of marine transportation fleet. *Ocean Industry* 13 (2):51–65.

Tucker, A. J. 1946a (12 August). Underwater drilling for any depth. *The Oil Weekly*, pp. 12:60–66.

———. 1946b (16 September). Bollard method: For shallow water drilling. *The Oil Weekly* 98–104.

Turner, R. Eugene and Donald R. Cahoon. 1988. *Causes of Wetlands Loss in the Coastal Central Gulf of Mexico*. Minerals Management Service: New Orleans

U.S. Department of Commerce, Bureau of the Census. 1850, 1860, 1870, 1880, 1890, 1900, 1910, 1920, 1930, 1940, 1950, 1960, 1970, 1980, 1990. *Census of Population*. Washington, DC: Government Printing Office.

U.S. Department of Commerce. 1959, 1964, 1969, 1974, 1979, 1984, 1989. *County Business Patterns*. Washington DC:Government Printing Office.

Veblen, Thorstein. 1934. *Theory of the Leisure Class*. New York: Modern Library.

Weeks, Lewis G. 1967 (April). Offshore operations around the world. *Offshore* pp. 41–59.

White, Brooks. 1993. *Statistical Abstract of Monroe County*. Marathon, FL: Monroe County Industrial Development Authority.

Williams, Neil. 1934 (31 May). Practicability of drilling unit on barges definitely established in Lake Barre, Louisiana tests. *Oil and Gas Journal* p. 47.

Wilson, Edward. 1982. MAGCRC: A classic model for state/federal communication and cooperation, pp. 72–90 in Joan Goldstein (ed.). *The Politics of Offshore Oil.* New York: Praeger.

Wise, David and Thomas B. Ross. 1954. *The Invisible Government.* New York: Random House.

Wolff, Paul. 1949 (August). Barnsdall-Hayward barge: A development for offshore drilling. *World Oil* pp. 87–96.

World Oil Staff. 1951 (September). U.S. Tidelands grab only the beginning. *World Oil* p. 73.

————. 1952 (June). The Tidelands issue and you. *World Oil* pp. 65–67.

————. 1953a (February). Cartel and Tidelands issue high on Eisenhower agenda. *World Oil* p. 46.

————. 1953b (September). Legal barriers are still holding up offshore exploration. *World Oil* p. 100.

————. 1953c (April). Continental Shelf could be a major resource for U.S. oil. *World Oil* pp. 64–65.

————. 1953d (June). Is this the answer to offshore operations? *World Oil* p. 116.

————. 1954a (October). Tidelands suits certain. *World Oil* pp. 107–108.

————. 1954b (August). Tidelands tiff denied. *World Oil* p. 98.

————. 1954c (April). Offshore work receives a boost. *World Oil* p. 72.

————. 1954d (June). Majors speed offshore efforts. *World Oil* pp. 95–100.

————. 1956a (April). Why Louisiana's offshore prospects look better. *World Oil* pp. 163–166.

————. 1956b (July). New platform has slanted piling for deepest Gulf drilling. *World Oil* pp. 116–117.

————. 1957a (November). Offshore drilling goes deeper. *World Oil* pp. 130–131.

————. 1957b (May). Transportation practices. *World Oil* pp. 122–123.

————. 1957c (May). Offshore picture. *World Oil* pp. 118–121.

————. 1957d (May). Drilling practices. *World Oil* pp. 132–147.

Wunnicke, Esther C. 1982. The challenge of the Alaskan OCS. pp. 91–102 in Joan Goldstein (ed.) *The Politics of Offshore Oil.* New York: Praeger.

Zondag, Cornelius H. 1952 (November). What's behind cartel charges? *World Oil* pp. 79–84.

Index